Escape from the USA

Escape from the USA

Erwin "Erv" Krause

© 2017 Erwin "Erv" Krause
All rights reserved.

ISBN-13: 9781546720447
ISBN-10: 1546720448

Contents

Prologue · vii

The Scenic Seaside Resort of College Point · · · · · · · · · · · · 1
Thoughts (Not So Deep) on Why We Travel · · · · · · · · · · · · 5
Even as a Youngster · 8
Bird-Watching…and Rock 'n' Roll? · · · · · · · · · · · · · · · · · 11
The *Gillette Cavalcade of Sports* · · · · · · · · · · · · · · · · · · · 15
A Pub Crawl of Epic Proportions · · · · · · · · · · · · · · · · · · · 18
Taking a Stand · 22
Harley on the Ropes · 26
About That '49 Harley · 32
About Those Vermont Plates · 39
My Friend, the Motorcycle Jacket · · · · · · · · · · · · · · · · · · 43
The Motorcycle Jacket Gets Me into Trouble · · · · · · · · · · 50
The South for the First Time · 55
The South for the Second Time · 58
Escapism · 64
All Come to Look for America (The Times, They Are
A-Changin') · 77
The Great American Road Trip · 82
The Departure (The South for the Third Time) · · · · · · · · 85
Barnyard Humor and Hospitality · · · · · · · · · · · · · · · · · · · 89
Outside Agitators · 92
The Mystique of Grits · 95

The Get-Out-of-Jail-Free Card · · · · · · · · · · · · · · · · · 98
Still Bird-Watching (on My Harley) · · · · · · · · · · · · · · ·104
The Fouke Monster ·109
Bury My Heart in Texarkana · · · · · · · · · · · · · · · · · · ·117
Arriving in Corpus Christi· ·123
"Corpus Christi" Means "Body of Christ" · · · · · · · · · · ·126
The Kindness of Strangers· ·140
The Episode ·145
The Way Home ·149
The Hill Climb ·153

Epilogue· ·163
About the Author ·169
For Your Listening Pleasure: Songs mentioned
(or just alluded to) in *Escape*· · · · · · · · · · · · · · · · · · ·171

Prologue

Is this a memoir? Hopefully not—too much of that going on these days! No effort here to share all the prosaic events of my entire life, for which you should be thankful. Perhaps it can best be described as a "slice of life," most of which took place during the summer of '66—in that magical decade.

It was also the year that I graduated from CUNY Queens and when I experienced a growing realization about how the rest of my life would unfold. Graduate school was not far off; I had already accepted an assistantship from the University of Illinois. Then would come a career. Perhaps I hoped to prolong my adolescence a bit before entering the "real world," whatever that was. I was ready for all that. But at the same time, I longed for a real adventure and realized that this might be my last shot. What would fulfill this desire better than the ultimate American experience: the road trip!

My friend Paddy's life had taken a different path. He'd been recently discharged from a stint in the army, and although he had been lucky enough to avoid Vietnam, he was now suffering PTSD of a different kind: from the fiery end of his marriage within months of the vows. I saw it as my duty to convince him to join me on the adventure. We would ride our Harleys from College Point, our hometown on the north shore of Queens, New York City, to Mexico. And back, of course—although there were moments when that return trip was in doubt. We were looking for adventure and would get our share.

In every person's life is a story or two worth telling, if they can do it properly, and I had been thinking about the summer of 1966 for a long time. And then, with little warning, along came the fiftieth anniversary of the adventure, something not to be taken lightly. A faded and tattered photo of my 1949 Harley resides in my wallet, where it's been for the past fifty years. Recently, every time I looked at that photo, the old bike seemed to say, "Go ahead and tell the damn story!"

The Scenic Seaside Resort of College Point

I've already mentioned College Point...the scenic seaside resort of College Point, that is, a peninsula bathed by the well-lubricated waters of Flushing Creek and a portion of the East River that mixed with the Long Island Sound. The place where my twin brother, Willie, and I grew up...our hometown. Only four roads connected "the Point" to neighboring Flushing and Whitestone and, to some extent, to the rest of the world. Three of those roads ran across swampland that was rapidly being encroached upon by landfills and auto-salvage yards—hence the derisive "Garbage Point" nickname. Fourteenth Avenue was the only "high" way, running across the glacial ridge, connecting the village with its neighbors. In the community's halcyon days, the late nineteenth and early twentieth centuries, that same avenue had been lined with mansions of prominent industrialists and professionals and formed a demographic dividing line between the affluent north side and the working-class area to the south. My parents' two-family home had straddled that "border."

Mom and Dad (Katie and Willie to most) had been German immigrants who fled the economic turmoil of their homeland in the 1920s to seek new lives in *Amerika*. Mom actually celebrated her sixteenth birthday aboard the passenger ship *Berlin* on December 14, 1926. My parents met here in America through mutual friends, but after I discovered in a Queens County telephone directory that our family name, Krause, was immediately preceded by mom's maiden name, Kraft, I would forever joke that their convergence must have had something to do with it: "You met each other through the phone book!" was the standard line. They married shortly after Mom turned twenty-one (Dad was twenty-six), and ten years later, by virtue of thrift and hard work, finally managed to purchase the attached brick two-family house at 12-34 116th Street. When I was a kid, Dad would repeatedly remind me that real estate was a wise investment and would always add another maxim: "...and when you do buy your first house, don't get a single-family...it's not a good investment."

One of my earliest memories is of plucking clothespins out of a cloth sack and handing them to my mother as she hung laundry out on the long clothesline suspended between my bedroom window and a metal pole in the backyard. All the homes on my block had garages in the rear and clotheslines suspended above the long, communal driveway that ran behind us. On sunny days, especially Saturdays, neighborhood laundry would sway about like Buddhist prayer flags. Back in the 1940s and '50s, washing machines weren't found in every home, and even fewer had dryers—Old Sol handled that chore!

Even first-time visitors could sense that the town's best days were somewhere in the rearview mirror, which longtime residents had known for some time. A melancholic "good ol' days" refrain held sway in much of their conversation. For new arrivals, however, who were fleeing other rapidly changing parts of the metropolitan area (i.e., "White flight"), the community seemed idyllic. "We thought we had moved to the country," many of them would say.

The area had been known as Strattonport back in the early nineteenth century until a certain Augustus Muhlenberg decided the location would be ideal for a new seminary. The institution itself didn't last very long, but the name "College Point" stuck.

Escape from the USA

Muhlenberg was a member of a prominent German-American family of clergymen/naturalists/scientists/military men dating back to the 1700s. Legend has it that back in 1776, John Peter Gabriel Muhlenberg, minister of a Lutheran congregation in Woodstock, Virginia, had been preaching a sermon based upon the third chapter of Ecclesiastes ("To everything there is a season...a time of war and a time of peace") and then declared, "...and this is a time of war!" He had then removed his clerical garb to reveal a colonel's uniform (he was an officer in the Virginia Military), and within half an hour, he had enlisted 160 men to join the rebel cause. JPG went on to serve admirably in our revolution and was instrumental in the defeat of the British in the final showdown at Yorktown. And then there's Muhlenberg College in Pennsylvania, named after the illustrious Henry Melchior Muhlenberg. And what about that rare turtle species *Glyptemys muhlenbergii*, commonly known as the bog turtle and familiar to all herpetologists? That little fellow is named after Gotthilf Heinrich Ernst of the same clan, a self-taught biologist and clergyman (not an unusual career path back in those days).

In the early part of the nineteenth century, the peninsula attracted many German immigrants, and by the 1850s, College Point's Roman Catholic, Lutheran, and Methodist churches were conducting services in both German and English. Then came another German who would figure prominently in the town's further development: industrialist Conrad Poppenhusen. Old Conrad purchased the rights to a rubber-processing method from Charles Goodyear and built his own plant in "the Point." Stimulated by government contracts to manufacture canteens and other military equipment for the Union Army during the Civil War, the enterprise thrived. A growing demand for factory workers was met with even more German immigrants. A benevolent capitalist, Poppenhusen provided housing for his workers and even established the first free public kindergarten in the United States for the children of the community.

College Point also became known for *garten*s of another variety. During the late 1800s, the town's large Teutonic presence, along with its leafy environs, resulted in a proliferation of biergartens and lesser drinking emporiums and even a brewery to satisfy the thirst of locals as well as visitors from the crowded tenements of Manhattan. By the turn of the nineteenth century, the community was thriving as both an industrial center and resort community, an atmosphere surviving well into the twentieth century. Even the Sultan of Swat himself, Babe Ruth, played on AHRCO Field (a block away from our house on 116th Street) during one of his barnstorming tours and reportedly hit one into the East River over five hundred feet from home plate—of course, it must have been high tide!

During my youth back in the '50s and '60s, College Point still maintained its Germanic atmosphere. There were many decent German bakeries, and I was an employee of one of them—Rupprecht's (later called Zach's)—during my high-school years. The bakers would arrive each morning at about 3:00 a.m., and my job consisted of cleaning, greasing, and buttering pans so they could go right to work. Many times, I would be finishing up just as they walked in the door!

There were numerous German butcher shops as well, providing residents with their favorite cuts of meat and wursts. Each had its own specialties, but if I had to name my favorite, it would be the liverwurst from Doerfleins. And I still recall the November hunting season, when many of the butchers would cut up the deer that residents had shot upstate and how these deer would proudly hang on display in front of the shops—now, that's something you don't see anymore!

Thoughts (Not So Deep) on Why We Travel

*They've got the urge for going, and they've
got the wings so they can fly.*

—Joni Mitchell

The *völkerwanderung*, also called the barbarian invasions (funny how the Romans, whose major source of entertainment was watching thousands of people get slaughtered in the Coliseum, referred to those northern tribes as "barbarians," but that's for another discussion), occurred between approximately AD 300 and 1000, in which there was a more or less continuous migration of Goths, Vandals, Suebi, and other Germanic tribes through Europe that filled the power vacuum left by the fall of the Roman Empire. And it really didn't stop there. Later on, when Europe got too crowded, these same people would continue their wanderings across the Atlantic to the Americas.

No wonder Westerners love to travel. We come from such a rich tradition! Not so much like other people. The Chinese used to be great travelers. Their great "treasure fleets," led by Admiral Zheng He, sailed throughout much of the world back in the fifteenth century. Traveling the high seas back then was a rather ballsy thing to do, so it's somewhat surprising that many of their sailors were eunuchs (ouch!), including the great admiral himself. Apparently, it was the adoption of Confucianism that helped foster the mantra, "There's no place like home, there's no place like home…" And, just like that, in 1433, the famed treasure fleet was scuttled—which explains why the Americas were not colonized by the Chinese. It also explains why the British, French, and other Europeans colonized China and not the other way around.

If you are wondering what any of this has to do with my motorcycle trip, I fully understand. I've been criticized at times for not being direct. You can either bear with me a little while longer or skip this part and turn to the next chapter; I promise I won't take it personally.

I see you've chosen to endure my apparent digression, so you might be saying to yourself, what about the Japanese? Aren't they inveterate travelers? Judging by the hordes of camera-toting Japanese tour groups found all over the world, your question would be understandable, but that's a relatively recent thing. One theory is that prior to the invention of the camera, visiting a foreign country would have made little or no sense to the average Japanese person. But it's a well-known fact that invention is the mother of necessity, and modern Japanese folks, now armed with their optics, went on to develop their infamous and insatiable desire for travel. Some say that if the camera had been invented in 1492, today's average American would be somewhat shorter in stature, and sake would be the beverage of choice, although our national pastime would still be baseball.

One of my favorite words is *wanderlust*. It doesn't just express a seemingly innate human desire to travel, to light out for the territories, but it goes even further to elevate that desire to the more sexual level: *lust*. And yet another travel-related word is an obscure biological term also borrowed from the German language: *Zugunruhe* (tzug-UN-ru-ah). Literally, it means "migration unrest" and describes the mass flocking and apparent restlessness of birds and other creatures prior to their annual migrations. If you're from the Northern latitudes, you may have witnessed those cacophonous gatherings of blackbirds and grackles in tall trees during the early fall—it's an avian form of wanderlust!

An interesting topic is travel. Whether the road trip (that Americans probably invented—and if not, certainly elevated to a form of high art) or something more sophisticated, is it possibly just a vestige of what our ancestors have been doing for millennia since that first great wandering out of Africa? Maybe we travel simply because we can! It explains everything, from why the chicken crossed the road to much of our human history. Many modern humans continue the travel ritual to this day. Perhaps it's symbolic and on a certain level pays homage to the völkerwanderungs of our forebears. Granted, there's a distinction to be made between voluntary wanderings and the migrations of those fleeing economic or political oppression at the hands of the Mongols, the Huns, the Cossacks, the Nazis, the Communists, or the US Cavalry. It is said that there's even a distinction to be made between traveler and tourist: the traveler is one who doesn't know where he or she is going, while the tourist is someone who doesn't know where he or she has been. If that's true, then on that motorcycle trip in 1966, my friend Paddy and I were travelers indeed!

Even as a Youngster

As a youngster, I had already developed a thirst for travel and a desire to explore new places. And if I could manage on my own, without adult supervision, all the better! At the age of eleven and twelve, my brother Willie and I, along with our friends, would plan Saturday "bike hikes." Packing knapsacks with lunch and tools and patch kits for inevitable flat tires, we would depart from College Point for destinations in Nassau and Suffolk Counties. These journeys would be instrumental in helping us become more independent, more grown-up, and in developing self-reliance and expanding our horizons, although at the time, we probably would have said we were just having fun.

I clearly recall the time Willie and I rode our "English racers" (a kind of catchall term for the thinner-tire, three-speed, hub-gear bike as opposed to the heavier, fat-tire, no-gear American Bike) to the Nassau County Fair in Roosevelt Field in Mineola, Long Island—which, incidentally, turned out to be the last event of its kind in Nassau: the county simply became too suburban. In addition to visiting the usual 4-H animal exhibits and eyeing displays of award-winning pies, we wandered over to the "freak show" area, perhaps lured by a sign proclaiming: "Prepare to be AMAZED by the TURTLE WOMAN from Georgia...half Human, half Reptile!" We entered that tent and exited, well, not too amazed. The "Turtle Woman," it turned out, just needed the services of a good dermatologist.

Escape from the USA

But the neighboring tent held out more promise: "Enter Here to see the Incredible She-He...Half Man, Half Woman... as Featured in *Ripley's Believe It or Not!*" We entered along with a small group of other perverted teenagers and young men, and before long, a skinny guy with a pencil mustache and a bad suit—really bad—introduced us to a large husky-voiced black person of indeterminate gender in loose-fitting garb. Everyone in the small audience got to see her face—one side covered in hair, the other side smooth. She (he?) then raised one billowy pant leg, revealing a large, hairy gam and then proceeded to show us the other, smooth and hairless—predictably. So far, totally *un*amazing!

Then the bad suit stepped up and made an offer to the crowd: for an extra fifty cents a head, the She-He would take us behind the curtain and reveal the more delicate parts of her anatomy to dispel any doubts regarding her amazing duality. Willie and I quickly reached into our dungaree pockets and scrounged up two quarters and a dime between the two of us. We instantly experienced pangs of regret over that lousy cotton candy we had wasted our money on a half hour earlier. We were left with no choice but the old, "Odds or evens—once, twice, three—shoot!" method, or did we flip a coin? Quite frankly, I've forgotten. But I do remember that my brother won, so I had to wait outside while he got to see the best part of the show!

Although it seemed like a long time, it was probably only five minutes later that Willie emerged from the back room, along with the other curiosity seekers, and I could tell from the generalized "We've been gypped" look on their faces that there was indeed a sucker born every minute—maybe even more than one.

Except for my brother, who either refused to admit that he'd been ripped off or maybe was in possession of better vision or just better imagination than the rest, because he immediately pulled me aside and whispered "You're not going to believe what you just missed!" and gleefully proceeded to regale me with a titillating description of all the anatomical details.

The ride home, with a dime to our names for an emergency phone call—just in case—was a miserable two-hour ordeal in a drenching rain. My mother took one look at us and immediately prescribed the usual remedy: a hot bath followed by hot tea with lemon juice. It worked!

Not long after the county-fair episode, we began a new school year, and my brother lost no time in sharing his experience with the other boys in our class, insisting that he had seen the real McCoy. Each retelling of the story was more embellished than the last.

So, isn't that why we travel: to learn about the world and expand our horizons? Ten years later, my friend Paddy and I would find ourselves continuing the travel tradition with an ambitious adventure, a motorcycle trip with a loose itinerary and vague destination—and part of me was hoping that the whole thing wouldn't turn out to be just another freak show!

Bird-Watching...and Rock 'n' Roll?

How is it that we develop seemingly idiosyncratic interests at an early age? Among the kids in my neighborhood, there were coin and stamp collectors. There were the comic-book aficionados. One kid in the grade above me was already a budding entomologist at the age of nine.

I was a bird watcher—not a "birder"; that term did not yet exist. My interest in the avian world started at about seven or eight years of age, and to this day, I'm at a loss to explain where it came from. I remember first purchasing relatively basic field guides and then graduating to Roger Tory Peterson's *Field Guide to the Birds*, which accompanied me on forays through the parks, woodlands, and marshes of the Queens neighborhood of College Point. Sadly, by the time my teenage years came to an end, so had the end come to Indian Rock Woods, along with the marshlands of Indian Valley and Powell's Cove, or what the kids called the "the Dirt Road." These surprising pockets of biodiversity would soon succumb to the rapid postwar development of suburbia, leaving only memories. But what memories!

Springtime was especially magical. Each April and May, I would spend hours observing the many neotropical migrants that frequented our woodlots. Many of these locations were called "empty lots." Hmm..."empty"? Really? These woods were filled with warblers in mating plumage: black-throated blues and black-throated greens, chestnut-sided, and the little black-and-white warblers that scampered up tree trunks like miniature woodpeckers. I recall being wonderstruck the time I saw a male scarlet tanager for the first time!

In late fall and winter, migratory ducks and other waterfowl would seek haven in the East River off Chisolm's Park, the tip of the peninsula that put the "point" in College Point. My visits to the park would start during the frigid predawn, when waterfowl were still relatively close to shore. Utilizing an old pair of binoculars and my Peterson's, and, of course, stealth, I was able to get close enough to identify them. Who besides me knew about the red-breasted and hooded mergansers, buffleheads, ruddy ducks, and the huge rafts of canvasbacks and scaups that thrived in the highly suspect waters right off our shore? Who else really cared?

So I was a bird watcher. But as I entered my teens, I realized that such an odd and nerdy pastime would not stead me well in the competitive climate of achieving and maintaining "coolness." At this point in my life, I still hadn't figured out that if you had to work at being "cool," you were never going to be. But I tried: by making sure to have cool dungarees and belt and shoes to match, and a modest collection of cool 45-rpm records—and, above all, by keeping that bird-watching thing under wraps.

Yet there was that one time, maybe the only time in my life, at about the age of thirteen, when my interest in birds was actually cool. It happened at a neighborhood birthday party. While hanging out in the living room, struggling to combat my awkwardness, I spotted a magnificent taxidermied cock pheasant perched on the fireplace mantel. I couldn't keep my eyes off it.

Sensing my obvious interest in the specimen, birthday girl Margie approached. "That bird?", she said. "My mother hates it! She's been on my dad's case to get rid of it."

"Are you kidding? I think it's cool! Tell your parents if they ever decide to get rid of it, I'll gladly take it!". Although I never went as far as actually admitting to be a bird watcher, I do recall being concerned that my obsession with the bird might arouse suspicion, so I dropped the subject.

It was also at this time, the midfifties, that a disc jockey by the name of Alan Freed started to become popular on New York City's 1010 AM radio station. Freed took the black music genre of rhythm and blues and renamed it "rock 'n' roll." His program would feature the harmonies of groups such as the Platters, the Del Vikings, and Frankie Lymon and the Teenagers—music that decades later would be called "doo-wop." Artists like Chuck Berry, Little Richard, and of course, Elvis would also be played. The entrepreneurial Freed also sponsored live-music shows on the stages of theaters such as the Brooklyn Paramount that attracted thousands of young people. The formula for the success of his radio show was simple: play the music that kids wanted to hear, and constantly emphasize that it was *their* music. It was also the attraction of forbidden fruit—it was "black" music, after all, no matter that many of the performers were white. And the not-so-subtle sexuality of the music concerned many parents: would their teenage children be led down the path of promiscuity? It was a moot point. It became our music and no small part of teenage rebellion.

Freed's radio program was largely a request show. In his distinctive rapid-fire delivery, he would introduce much of the music in a highly personalized manner: "…and the next song goes out to Tony of Flatbush from Angela, who have been going steady three months, two weeks, and one day: 'Sincerely,' by the Moonglows."

I was one of Alan Freed's biggest fans, but at this age, I was uncertain about how my parents would feel about me listening to "black" music—while doing homework, no less. But I was addicted. So I kept my little Emerson transistor radio at a low volume, trying my best to stay off the parental radar screen. One night, about a week or so after that little neighborhood soiree, the one with the pheasant on the mantel, I was involved in my nightly ritual of homework and rock 'n' roll. Imagine my astonishment when Freed announced, "…and this next song goes out to Erv in College Point from Margie. 'Sorry, Erv, you can't have the pheasant…my parents decided to keep it!'"

The requested song? "Rockin' Robin," by Bobby Day. "He out-bopped the buzzard and the oriole!" Was that not cool?

The Gillette Cavalcade of Sports

To look sharp every time you shave
To feel sharp and be on the ball
To be sharp use Gillette Blue blades
For the quickest, slickest shave of all!

Every Friday night, this little jingle set to a robust march and the round-by-round commentary of announcer Don Dunphy in his distinctive lilting, Irish voice were eagerly anticipated by millions of boxing fans during the 1950s. These included Dad and myself. I'm talking about the *Gillette Cavalcade of Sports* (less formally, "the Friday-night fights"), a weekly event that brought the world's best pugilistic talent into America's living rooms during the early days of TV. For many fans, it signaled the unofficial kickoff to the weekend.

In the fifties, we were still in boxing's golden age (though perhaps also its twilight years). Rocky Marciano, Sugar Ray Robinson, Archie Moore, Gene Fullmer, Joey Giardello, and others were household names. For sheer athleticism, nobody could match Robinson. Smooth as silk, his marcelled hair remained unruffled, even by the end of the contest. Being left-handed myself, I paid close attention to southpaws like middleweight Carmen Basilio and attempted to emulate his ringmanship when sparring with friends.

Erwin "Erv" Krause

In 1960, an eighteen-year-old from Louisville, Kentucky, named Cassius Clay won the Olympic Gold Medal in boxing's heavyweight division, defeating his Polish opponent, Zbigniew Pietrzykowski, by unanimous decision. He then turned pro and burst onto the professional scene not only by recording thirty-one straight wins but brashly reciting poetry in which he would predict, with uncanny accuracy, the exact round in which a bout would end. His antics generated considerable controversy during an era in which humility was the hallmark of professional athletes, but they undeniably aroused considerable interest in his fights. Throughout the early part of his career, the "Louisville Lip," as he was called, had legions of supporters as well as detractors. I counted myself among his detractors, finding his braggadocio and his penchant for demeaning and taunting opponents offensive. His conversion to Islam and name change to Muhammad Ali certainly didn't help matters. My limited knowledge of that religion informed me that Muslims had not only sanctioned slavery historically, just as Christians had, but in certain parts of the Islamic world, slavery was still practiced!

Every time Cassius Clay stepped into the ring, I would root for his opponent. In June of 1963, my wish almost came true when the British champ Henry Cooper nearly knocked him out with a devastating left hook toward the end of round four—but Clay survived. The fight was stopped by the referee in the next round because of a gaping cut over Cooper's eye.

The most controversial fights—and most bizarre—were Ali's fights with Sonny Liston, whom Ali derisively called the "big, ugly bear." The first fight ended when champion Liston refused to come out for the seventh round, and the championship was awarded to Ali. A rematch was held in 1965 in Lewiston, Maine, the first and only heavyweight championship bout to take place in the Pine Tree State. To this day, I remain skeptical regarding the first-round knockout scored by Ali with his infamous "phantom punch." Having seen replays of that fight many times, I'm convinced that still hidden out there somewhere is another explanation for Liston's loss.

During my teen years, my friends and I converted my basement to a workout room. Barbells and dumbbells were acquired either by barter or bequeathed to us by those who had lost the commitment. There were also boxing headgear and gloves, and a speed bag suspended from the wall. We would occasionally put a mat down, put on the gloves, and engage in brief bouts that ended not by knockdown or decision but in another bout—of laughter. Usually, however, the speed bag would be my opponent, and I would work on snapping right-hand jabs (remember, I'm a southpaw) followed by potent left crosses—potent in my mind, at least. In school yards and occasionally on the street, we would "slap fight" using open hands so as not to really hurt one another, but I recall delivering and receiving some really good shots!

There were, of course, real fights. They took place after school, usually in the school yard, in the park, or in empty lots. Later on, in my late teens and early twenties, barrooms would be the most likely venue. More spontaneous clashes, those sudden physical eruptions of youthful anger, would usually be broken up quite quickly by friends. Planned fisticuffs ("I'll meet you in the school yard at three o'clock!") were allowed to continue until one of the combatants emerged victorious. Over the course of many years and many fights, not once did I ever witness a knife or any weapon brandished.

A Pub Crawl of Epic Proportions

Some degree of misspent youth is a necessary rite of passage to a reasonably honorable adulthood. In that spirit, allow me to return to the subject of College Point's drinking establishments. By the 1960s, only two of the town's original dozen or more biergartens—real ones, that is—remained: the Village Grove up on Fourteenth Avenue near Whitestone and Eifler's, which later became Flessel's, on Fourteenth Road and 119th Street. Flessel's had an ornate mahogany bar with a wonderful onyx rail that was not only ergonomically functional but an aesthetic delight for those who appreciate such things. That establishment also boasted enormous porcelain urinals in the men's room dating back to the 1800s. Family lore has it that as a four-year-old, I was accompanied by my father into this bathroom, and as I stood before the urinal, I looked up and remarked, "Dad, am I here to take a shower or take a leak?"

Numerous other drinking establishments lined the Main Street and graced many street corners in more residential areas as well. These watering holes served a discriminating clientele often until 4:00 a.m. (actually, they often stayed open even later, since although patrons couldn't be served after four, they could remain to finish the drinks already at the bar and often exercised that legal right) before reopening at 8:00 a.m., and astonishingly, there would be regulars seated on their stools at 8:01 a.m.! Whenever a College Pointer was asked if it wasn't too early to be drinking, his justification was, "Well, it's eight in the morning somewhere in the world!"

Upon reaching my eighteenth birthday in 1961 (the legal drinking age in New York back in those enlightened times), my friend George made me an offer that I was too polite to refuse: a beer at every bar in College Point...on him!

We strategized. The undertaking required a plan, and so we dutifully mapped out a route, even though the Point wasn't much more than a square mile or so. The odyssey would begin by car. George's old 1951 flathead Ford was reliable enough to handle the first leg, which included some establishments on the periphery such as the Village Grove and Behan's out near Tallman Island. At a designated halfway point, we would responsibly park the car and continue on foot to the many saloons clustered around Main Street. The planning took place on my actual birthday, Thursday, November 30, but the tour itself would take place the following evening.

That Friday evening, my mother served George and me a well-crafted culinary creation. I think it was goulash. Mom loved to feed my friends, and if they happened to be around when supper was being prepared, there was a standing invitation. She also knew about the hazards of imbibing on an empty stomach. Aware that we were dressed up a bit more than usual, she inquired as to our plans for the evening. We gave an evasive, "Nothing much...just goin' out."

She knew better. "Don't drink so much...and call if you're going to be late" was her final admonition.

In many of the bars, George announced that we were celebrating a momentous milestone. Being of Scotch ancestry, he was hoping to coax a free beer out of the bartender. (At fifteen cents a beer, the going rate in 1961, and with a few free beers thrown in, George's tab for me still came in under five dollars.) In some bars, however, discretion dictated that the birthday aspect not be mentioned, since I had already been an occasional patron during the past year—bringing attention to my age might have proved awkward. Not bringing it up was especially important at the G&W, a place where we had become almost regulars since the age of sixteen after hard games of school-yard basketball at PS 27 just up the block.

I believe I've forgotten to mention something, something very significant to the telling of this story: the challenge facing us that night involved thirty-three drinking establishments. That is correct. A total of thirty-three public houses, bars, saloons, dives, brauhauses, lounges, taverns, gin mills, beer parlors, and two genuine biergartens served the needs of the good burghers of College Point in 1961, and we did not plan to miss a one. An astonishing number indeed for such a small community. And, although likely apocryphal, a story circulated about town that, according to that unassailable source of useless trivia (no, not Google…that didn't exist until almost forty years later) *Ripley's Believe It or Not!*, the Point had the highest number of bars per capita in the United States—or was it in the entire world? Could it be that our epic "pub crawl" (we referred to it as "bar-hopping") also deserves some recognition in Ripley's? Maybe it's still not too late!

But I already hear the voices of skeptics. They're doing their little calculations and coming to the conclusion that consuming such an amount of alcohol in the course of a single evening would result in acute alcohol poisoning—or worse! Dear reader, I assure you—I do not lie. Exaggerate, perhaps. But before you pass judgment, allow me to explain.

This much is fact: I was served thirty-three regular American draft beers. I should note that so-called light beer, with its somewhat lower alcohol content and insipid taste, had (thankfully) not yet been foisted upon the public. On the other hand, the typical beer glass back then was smaller (maybe eight ounces), and some bartenders had stopped using those ubiquitous plastic scrapers supplied by beer companies, which resulted in disturbingly large, foamy heads that would quickly evaporate, leaving rather meager contents. And I never said that I drank an entire glass, especially after the twentieth stop at a dive known as the Irish House, where it was becoming apparent that discretion was indeed the better part of valor and that some sort of temperance on my part would be necessary for me to complete my mission: to reach bar number thirty-three (which happened to be the bowling alley where my cousin Harry tended bar on weekends). By the way, if you're into numerology, 1933 also happened to be the year that the Twenty-First Amendment, the repeal of the Volstead Act, took place to end Prohibition—something that wasn't on my mind then but is now.

I'm not exactly proud of leaving out the part about those unfinished beers when I tell this story, but as a friend once advised, "Never let the truth get in the way of a good story!"

Taking a Stand

It was at the age of seventeen in 1961 that I first began to question America's foreign policies. Maybe it was the Bay of Pigs invasion, that ill-fated, CIA-sponsored attempt to wrest Cuba back into its proper state of compliancy with its benevolent big neighbor to the north.

I had learned that the pro-American dictator Fulgencio Batista, who had come to power via military coup in 1952, controlled the island as a despot and had developed close personal ties to Mafioso guys like Meyer Lansky and Lucky Luciano, enriching himself at the expense of the people. So, when Fidel Castro and his boys finally came out of the mountains and overthrew Battista, I thought, *How can this be a bad thing?*

But President John F. Kennedy and his advisors had concerns about the left-leaning politics of the new leader and immediately began formulating strategies to overthrow him. The Bay of Pigs invasion, involving CIA-trained Cuban émigrés, was launched and thwarted within the same twenty-four hours. Our intelligence people had claimed that thousands of freedom-loving Cubans would rise up to support the force and ensure the plan's success. Instead, the mission proved an embarrassing failure and established a new standard in the debacle category. Although our government at first denied involvement, the truth soon came out, delivering a major blow to American prestige during the Cold War. On a personal level, the event gave me grave doubts about the wisdom and morality of our foreign policy.

Then came Southeast Asia. Beginning in the nineteenth century, the French had begun to assert themselves in this region, eventually creating the wonderful colony of French Indochina, still renowned for its legacy of fusion cuisine. But it takes more than great food to maintain a relationship, and by the early part of the twentieth century, a movement against French colonialism began to intensify. During World War II, the Vietnamese made it clear to the Japanese that they too were not welcome, and (can you blame them?) the Vietnamese people expected to be granted complete independence after the war. After all, wasn't freedom what the Allies had fought for? But the French stubbornly attempted to maintain control of their colony until being decisively defeated at the Battle of Dien Bien Phu by Vietnamese forces under the leadership of Ho Chi Minh, and a two-state solution, resulting in North and South Vietnam, was agreed upon.

Concerned that South Vietnam and perhaps all of Southeast Asia would fall to communism (the domino theory), the United States soon began to send military advisors to assist the pro-Western South Vietnamese government against pro-communist Viet Cong rebels being supplied by the North. First Eisenhower, then Kennedy and Johnson, continued to ramp up US troop deployment, and by the middle of the 1960s, we were involved in a war costing thousands of lives and fanning the growth of an antiwar movement at home. Our country was clearly divided.

My own readings about the history of Southeast Asia combined with a growing distrust of our foreign policy gradually led to my conviction that our actions in Vietnam were at best misguided. Then came information about our use of napalm and Agent Orange and the unsettling accounts of atrocities against civilians. I began to view our pursuit of the war as beyond misguided and toward clearly evil. Our country itself was becoming increasingly divided over the issue. This division could be seen in all aspects of American culture, including popular music. In the midst of a growing antiwar movement, "The Ballad of the Green Berets," by Staff Sergeant Barry Sadler, became a rallying anthem for the prowar side and a number-one pop-music hit. In 1966, Barry McGuire (another Barry!) recorded the antiwar anthem "Eve of Destruction," which also made it to number one on the charts.

Living in a largely conservative, blue-collar community, I didn't always find it easy to express my growing antiwar sentiment. The attitude of many was summed up in slogans such as, "America: love it or leave it!" or, "If your heart is not in America, get your ass out!" The criticism that hurt the most was that the antiwar movement supported the enemy and thus demoralized our troops and even cost American lives. I found myself trapped in that ultimate paradox faced by a citizen of a democracy: exercising one's right—no, one's *duty*—to make informed decisions but then being labeled "unpatriotic" when taking a stand contrary to the government's agenda. I was seen by many as un-American, especially by the "Our country, right or wrong!" crowd. I found support in the words of Carl Schurtz, a German-born Union general during our Civil War and a statesman who was famously credited with saying that true patriotism is not "our country, right or wrong" but rather, "our country, when right, to be kept right; when wrong, to be put right."

As I've mentioned, I was an avid boxing fan but also a major disparager of Cassius Clay (aka Muhammad Ali). But something happened to change my opinion in the spring of 1966. It was at this time that Ali first publicly voiced his disagreement with America's involvement in Vietnam when he famously told

Escape from the USA

reporters, "Man, I ain't got no quarrel with them Viet Cong," and, more to the point, "No Viet Cong ever called me 'nigger!'" These remarks were the first salvo in a flurry of jabs he directed against our nation's role in that war, and they led to his refusal to be drafted into the army, followed by him being stripped of the World Heavyweight title. It was also the first time I finally mustered up respect for him as a person.

I found myself struggling with a difficult decision: if I got drafted, would I leave the United States for Canada? I vividly recall my father's reaction when I discussed my dilemma with him. Dad not only promised me support but applauded my position, saying, "I lost my father and my only brother to war, and I'll be damned if I'm going to lose a son...especially in this stupid war. If you decide to leave the country, I'll help you in any way I can!"

As it turned out, I never had to make that decision. Although I was drafted (in those days, New York City residents got letters from the draft board along with a clearly visible subway token—a free ride to the preinduction-physical facility on Whitehall Street in Lower Manhattan) and reported for a physical, which I passed, I was granted a deferment as a full-time graduate student at the University of Illinois, followed by additional deferments upon entering the teaching profession. This may explain why I'm still a citizen of the United States and not our friendly neighbor to the North.

Harley on the Ropes

I promised a story about a road trip, and it's time I steered in that direction. A road trip requires a vehicle, which by definition is "a means of transmission or transport." But clearly, there's more to it. Think of the myriad songs, movies, and stories extolling the adventure of the open road and the cars and motorcycles that make it possible. Think of Chevy's iconic advertising jingle, "See the USA in Your Chevrolet…" Like many Americans, my fascination with cars and motorcycles began early in life. As for cars, as much as I loved them and recognized their potential to fulfill my desire to experience the country, actually owning one during my late teens and early twenties was a financial impossibility.

For a brief period, I tried car ownership—shared ownership, that is. My brother and I went in on a '49 Ford Coupe powered by that same legendary flathead V-8 that old Henry Ford first put into his Model A back in '32. A quirky car by today's standards for sure (I recall the vacuum-driven windshield wipers that would just about quit on you whenever you pressed the accelerator…always just when you needed them the most!), it was reliable for its time, with the added advantage of a cavernous trunk. It was called a "business coupe" because that huge trunk enabled traveling salesmen to haul their wares. It held four or five spare tires for the inevitable flat, not to mention a case of motor oil to satisfy the motor's prodigious consumption.

Escape from the USA

But it was an experience I don't regret. As a rite of passage (not to mention its educational value), few things compare to the ownership of a second-hand Ford. It was my first car, after all, and you know how it is with a first love.

But "sharing" a car with my brother wasn't exactly the best arrangement, I soon found out, and the insurance costs proved unmanageable. How dare the insurance industry label seventeen-to-twenty-year-old boys "high risk"! So we wound up selling it for a little bit more than the fifty bucks we'd bought it for, and I wound up borrowing my dad's car or a friend's when going out on a date.

But I still wanted a vehicle of my own, and the idea of owning a motorcycle soon became more and more appealing. Barring snow on the highway, I could still ride most of the year, and there was no denying the fun and adventure of two-wheeled travel. But the real advantage of a motorcycle (versus a four-wheeled vehicle) was the absence of an insurance requirement in many states, a fact that soon took on greater significance for me. The motorcycle love of my life was my first Harley—more about that machine and my relationship with it after some historical background.

With all the motorcycles on the road today, it's hard to imagine a time when the industry was in trouble, and this was especially true for Harley-Davidson. Factory figures for HD show a gradual rise in production during the postwar period, from 1,430 units built in 1946 to 12,924 motorcycles leaving the factory in 1949. A gradual decline in production followed until it reached a low point of 4,757 machines built in 1954. Production in the late 1950s and throughout the 1960s continued to be a bit up and down. The true low point came in 1969, when American Machine and Foundry (AMF), better known for its bowling equipment, bought the company, which turned out to be a true gutter ball for the venerable Harley-Davidson brand. Quality and sales declined considerably, and the company nearly went bankrupt. But in 1981, a group of investors including Willie G. Davidson (grandson of the founder) bought the company back from AMF, and a turning point was reached. Adopting a marketing approach extolling the retro and an appeal to buy American, the company became profitable, and production increased to historical highs. From the pathetically low production of the mid-1950s, fifty years later, Harley was manufacturing three hundred thousand "big twins" a year!

But back in the early 1960s, when I bought my 1949 Harley for $400 from a young guy in Brooklyn (whose new bride had issued the "It's either me or the bike" ultimatum), the company was in trouble. Paddy and I (he owned a '52) had good reason to believe that the company's days were numbered, and we envisioned ourselves as heroic keepers of the flame.

Another factor working against Harley in the '50s and '60s was increasing competition from English bikes such as Nortons, Triumphs, and BSA 650s, which were cheap and fast, despite being scorned by Harley owners. And if you were a touring kind of guy, you couldn't match the reliability of the drive-shaft, opposed-cylinder design of the BMW R90. Then came the Japanese motorcycles, inexpensive and highly reliable. The first one that got my attention was the Honda CB450 with its dual-overhead-cam engine. And the incredible four-cylinder 750-cc models had yet to appear.

There was yet another issue. The motorcycle industry, and especially Harley-Davidson, had developed a growing image problem. Stories about gangs such as the Hell's Angels and their violent rampages began to appear in the press, and Hollywood soon capitalized with exploitation movies such as the 1953 cult classic *The Wild One*, starring Marlon Brando and Lee Marvin. (Brando, by the way, did not ride a Harley in the movie.) Increasingly, motorcycle riders were viewed as troublemakers. I personally recall being told by a bartender to not park my motorcycle in front of his establishment, implying that it would be bad for business. On another occasion, I pulled into an upstate New York gas station to tighten a loose bolt on my motorcycle, and the owner came running out, threatening me with a wrench and demanding that I leave the premises. This time, I took a stand (my girlfriend was riding with me). I refused to leave, pointing to the sign that said "service station" while reminding him of the term's definition and that if he didn't back off—and so on and so forth.

But the image problem was a real one. In response to a notorious Hell's Angels incident in Hollister, California, the American Motorcycle Association (AMA) issued a public-relations release claiming that 99 percent of motorcyclists were law-abiding citizens. Almost immediately, motorcycle gang members began to proudly identify themselves as "1 percenters," many even wearing a small patch to indicate membership in this "club." Whether 1 percent or not, in the view of many Americans, most motorcyclists were troublemakers.

To deal with the problem, Honda's marketing people came up with an advertising campaign featuring the slogan, "You meet the nicest people on a Honda." The creators, Grey Advertising, developed a series of TV and magazine ads featuring wholesome members of society (young couples, housewives, civic-minded types, and such) scooting around suburbia on their little Honda 50s. The campaign's greatest coup came when American Honda sponsored the annual Academy Awards TV broadcast in 1964, and millions of viewers saw two ninety-second commercials each ending with that slogan.

Hollywood, on the other hand, wasn't quite finished exploiting motorcycling's "bad boy" image. The film industry continued the practice through the 1960s with the release of movies such as *These Are the Damned* (1963), *Motorpsycho* (1965), and *The Wild Angels* (1966), starring Peter Fonda. (I'm hoping that maybe the 2007 John Travolta movie *Wild Hogs* served to euthanize the genre for good!) As a general rule, these movies were terrible, with the sole redeeming value of being laughable. But, as with most things in life, there's always an exception to the rule.

In 1969, three years after our road trip (don't worry, I'm getting to it!), I was listening to Jonathan Schwartz, an iconic disc jockey of the 1960s, on radio station WNEW-FM. Schwartz would frequently intersperse music with social commentary and media criticism, and during this particular broadcast, he urged listeners to see a new movie called *Easy Rider*, which starred Peter Fonda and the relatively unknown Jack Nicholson. At first opportunity, I went to a theater on the Upper East Side of Manhattan to see it for myself. The final scene, where Fonda and Nicholson are blown off their bikes by pickup-truck-driving, shotgun-toting rednecks left me sitting speechless long after the entire audience had departed. No one, I thought—*no one*—there could possibly have related to that movie and the utter vulnerability of two riders in hostile territory the way that I could. In my lifetime, few movies could ever truly speak to me on the same emotional level as *Easy Rider* did. My reaction will become understandable as you continue reading.

Postscript

The impact of the motorcycle on the American cultural scene goes well beyond that wrought by the music and film industries. On June 26, 1998, an exhibit called *The Art of the Motorcycle* opened at New York's famous Solomon R. Guggenheim Museum. The show featured 114 motorcycles chosen for historic importance and design excellence. Since architectural critics have often compared the Guggenheim on New York's Museum Row to a parking lot, the venue was perfectly suited to it.

I was one of the many thousands in attendance gleefully marveling at the beauty of the collection, which went back to the turn of the last century to the present: early Harleys, art-deco Indians, classic BMWs, Vincents, and even bikes I had never heard of, like the Scott Squirrel Sprint and the Ducati M900 Monster.

With a total attendance of 301,037 in just under three months, the show attracted the largest crowds in the Guggenheim's history although some of the stodgier attendees could be heard muttering, "Yes...but is it art?" Art critic Jeremy Parker saw something more. He described it as an event that finally brought to an end the era of demonization and denigration of motorcyclists as social outcasts. To this day, though, I'm not sure that Parker's conclusion is shared by all.

About That '49 Harley

Down on the left side of the engine case, above the ignition timing-inspection plug, was engraved 49———, my bike's serial number. I suppose it could be argued that the machine was a '49, but to what degree was speculation. Very few fifteen-year-old Harleys (either because of attrition or design) were still truly "original." On my motorcycle, for example, the original hand-shift/foot-clutch mechanism (better known as a "suicide shifter") had already been replaced by a more up-to-date foot-shift/hand-clutch system. Neither were the gas tanks original (Harley had a two-tank design); something I discovered only after visiting the Harley-Davidson Museum in Milwaukee many years later. Other changes and updates had also been performed over the years.

That FL model, the flagship of the Harley line, had first been introduced in 1941. In 1948, the company had introduced a redesigned "Panhead" engine (so called because of the appearance of the valve covers) with an overhead valve/pushrod design, self-adjusting hydraulic lifters, and aluminum-alloy cylinder heads. It replaced the earlier "flatheads" and "knuckleheads," and it would remain in production with only minor changes for fifteen years. The engine had a displacement of seventy-four cubic inches (1,210 cc) and was rated at forty-eight horsepower. Not too many riders back in this era of cheap gas concerned themselves with fuel economy, but as one who enjoyed road trips; I estimated about forty-five miles per gallon, or about 150 miles on a tank.

The bike had a wishbone frame, meaning that it had no rear springs or shock absorbers. This was commonly referred to as a "rigid" frame. The only rear "suspension," if it could be called that, was the saddle-post spring. However, it did have the hydraulic front-end suspension that was introduced in 1949 to replace the older spring forks, and hence this model was designated the Hydra-Glide. The basic design would last until 1958, when a new frame featuring a rear swingarm with coil-over-shock suspension units was introduced, and the Hydra-Glide became a Duo-Glide. My friend and I composed a little ode to that "new" design:

> Ride in pain with a rigid frame
> Slip and slide with a Duo-Glide

It wasn't until 1965 that another radical innovation came along: electric starting! The new model was called the Electra-Glide. Up until then, Harley riders like myself employed the method—or, should I say, "ritual"—of kick-starting. Each machine had its own idiosyncrasies, and this was certainly true of the engine-starting process as well. To choke or not to choke? If to choke, half or full? Ambient temperature played a role, and so did whether the engine was already warmed up or if it was a cold start. Take my bike, for example. When running well (which, thankfully, was most of the time), it would start on the first kick (after a few priming kicks), or sometimes after the second or, rarely, a third—a thing of beauty. If not by the third, however, then at least nine or ten would be necessary—nothing in between—with possible carburetor machinations thrown in for good measure (not surprisingly, many older Harley riders had at least one bad knee!) In my frustration, I would make some reasonable requests of the engine, which might lead to pleas and, if that didn't work, some cusses, blasphemy, and gender-specific invectives. Don't judge me for these moral lapses: I was just doing what I thought necessary.

These machines were quirky right out of the factory. And let's not forget that they were manufactured in the beer capital of our nation: Milwaukee, Wisconsin. So, on Monday mornings, what sort of commitment to precision would be mustered on the assembly line?

And, how quirky? Here's some advice right out of the company's official service manual:

> With *some* engines, depending upon carburetor adjustment, hot starting is more dependable if the starter is given one stroke before turning ignition switch ON.

Or how about:

> Develop the habit of *frequently snapping the throttle shut for an instant* when running at high speed. This draws additional lubrication to pistons and cylinders and helps cooling.

Speaking of lubrication, those older Harleys did use oil—lots of it! Some was lost to seepage from valve-cover gaskets and other engine seams. Additional oil was diverted to lubricate the primary chain between the engine and tranny, and it then exited through a small opening in the chain cover. I used a small pan to collect the drippings and returned it to the oil tank. I always carried a couple of small aluminum pans along to avoid telltale puddles when parked in other people's driveways.

Escape from the USA

So what sort of Harley did a person want? A rigid-frame chopper with its impossibly miniaturized fuel tank might have looked real cool, but even a brief excursion through the minefield of potholes called 1960s New York City could have delighted only the Marquis de Sade. On the other end of the spectrum were full-dress Rococo creations favored by many black riders back in the day. Cleaning these garish machines could take hours, and to really detail them, you needed a full working day plus overtime, not to mention a gallon of Noxon Chrome Polish. Even then, the quality of the results would depend upon the eye of the beholder, who might express unabashed derision, a hymn of joy, or simply a look of bemusement.

There was also a number of what are now called "rat bikes," although that term was not a part of motorcycle parlance in the sixties. If they were automobiles, we'd call them jalopies. Back in the day, we called such a motorcycle a "barouche"— of whose etymology I plead ignorance. Easily identified by the well-advanced oxidation of exposed metal surfaces and a lusterless paint job or gray prime coat, these machines still had to run fairly well in order to have any cred. Today's rat bikes have been elevated to a cult-like art form, and owners take delight in the irony of the decent-running engines powering such sorry-looking bikes. These are superb examples of *wabi sabi*, or neglect by design, and demand recognition as such.

DE GUSTIBUS NON EST DISPUTANDUM

When undertaking restoration of my 1949 Harley, I had a plan informed by aesthetics and functionality. As for function, the machine had to be capable of transportation to the many far and near places I dreamed of. Aesthetics were governed by German-Lutheran requirements for modesty and simplicity and a deeply held conviction that beauty could be found in the understated elegance of rich, black lacquer against chrome. At least in theory. Fiduciary realities forced me into settling for a passable enamel finish (the by-product of the painting of an old Ford at my friend's body-and-fender shop) and some compromises in the chrome department. My neighbor Bob down the block was a big help. He owned a metal-plating business in Flushing and would frequently check in on my restoration project. "Why don't you give me some of those parts, and I'll plate them for you!" he finally volunteered. He was kind enough to perform the metallurgical services gratis. The end results were commensurate with his fees.

I was also lucky to have friends who were mechanical geniuses—the kind of guys who spent their high-school years walking around the halls with nothing but a single black-and-white-marble composition book but had every single engine part memorized by the time they were sixteen. Without them, I would never have undertaken the task of rebuilding a motorcycle.

As the bike neared completion, I had the smug satisfaction of owning a Harley that elicited occasional compliments. ("Erv, she looks great...what a difference!" Or, "I can't believe it's the same bike...") Hey, what are friends for? Some of those parts sent off to be "chromed" might have come back with luster slightly beyond that of a galvanized can, but I was delighted by the before-and-after transformation. At least the pitted chrome was gone. I had this thing about pitted chrome: to me, it was a like a bad case of pimples on a girl with an otherwise pretty face. The bike looked presentable, but I always considered it a work in progress.

At any Harley gathering, one immediately senses the "no two are alike" theme and that each machine reflects its owner's personality. A bike with that right-out-of-the-showroom look is an anomaly. This was certainly true in the '60s and remains so even

Escape from the USA

to this day: every bike is an expression of personal creativity. Not only manufactured *in* America but made *for* Americans and their cherished birthright of rebelliousness and individuality.

Some final thoughts: how often does the word "empathy" come up in the context of one's relationship with an internal combustion machine? That is exactly what developed between me and my Harley. I constantly checked the exhaust-pipe color (too dark...too white?) and fiddled with the carburetor's high-speed adjustment screw to achieve the perfect mix and assure the engine's happiness. I developed an awareness of my motor's rpm "sweet spot" when cruising down the highway. I had helped install those main bearings and connecting rods, after all. I was straddling that engine...wasn't it always talking to me, and wasn't I listening? Listening to the conversation between crankshaft and connecting rods, between pistons and cylinders, between valves and camshaft via pushrods and rocker arms...a rhythmic symphony of purpose. And, of course, coming to understand that just as relationships with friends or lovers are never perfect, so it is with your motorcycle. To make the whole thing work, you need to accept—nay, *embrace*—your motorcycle's peculiarities!

Perhaps most motorcycle riders never achieve that state of grace: "being one with your machine." Robert Pirsig addressed the dichotomy of the Romantic and the Classicist in his *Zen and the Art of Motorcycle Maintenance.* Thinking about this, I realized that I leaned more to the Romantic and had to work hard at developing the Classicist within me. I've read that book several times and am still uncovering its many rich layers of meaning. Pirsig had the kind of empathy I'm talking about.

Ultimately, it must be said that the relationship between man and machine is not something all people can wrap their heads around. You either get it, or you don't. Celebrated English singer-songwriter Richard Thompson obviously gets it in his iconic paean "1952 Vincent Black Lightning," released in 1992. The song's storyline goes something like this: desperado James, the proud owner of a '52 Vincent, has been eyeing a local lass, beautiful Red Molly, for some time now, unbeknown to her. One fateful day, Red Molly compliments James on his motorcycle.

James replies with a compliment to her good taste: "My hat's off to you. It's a Vincent Black Lightning, 1952." And so begins their tragic (of course…it's folk music!) and short-lived love affair.

The police soon catch up with desperado James, and in his last moments, he calls Molly to his "dying bedside" and, in an ultimate act of love, bequeaths the motorcycle to Molly——before a final kiss.

The real hero of the tale is the legendary 1952 Vincent Black Lightning motorcycle: what is it about that bike that inspires Thompson (via the outlaw James) to imbue an insentient object with *soul*? What is this mysterious quality, the soul? It's different things to different people for sure, but personally, I believe that when one finds soul in something like a motorcycle, it can only be the projection of one's inner soul.

You've got to hand it to the Brits for giving us a songwriter capable of writing what is, in my opinion, the ultimate motorcycle song. The best thing we Americans could come up with until then was the sophomoric "Leader of the Pack," by the Shangri-Las. When it came out in 1965, that song was all the evidence I needed that rock music's end days had finally arrived, and I was listening to its funeral dirge.

It wasn't until 1991 that Neil Young (ah, but he's Canadian!) gave us "Unknown Legend" (1991) with its soulful imagery of a blond haired lady riding her Harley along a desert highway. I never get tired of that song!

About Those Vermont Plates

Youthful transgressions and errors of judgment are part of that life stage from which only the lucky few emerge unscathed. Let me share with you the circumstances that required me to change my residency from New York to Vermont. Actually, such a change in residency was part of the master plan from the start, since back in the '60s, for a person of my age to register a motorcycle in New York State, with its onerous mandatory liability-insurance requirement, would have cost over $400 per year—even more than I'd paid for the bike! The motor-vehicle code of neighboring Vermont, on the other hand, had no such requirement. Remember, I was a college student working only part time, and such a larcenous expense would not have been possible. So, a plan had already been in place, but it was a certain incident that forced my hand.

Back in '62, I was working on my first real motorcycle project, the resurrection of the '49 Harley, and finally got the bike up and running. The motor sounded decent, and I was eager to take her out on the road. Problem was, I hadn't gotten around to the registration process. So, I had an unregistered vehicle with no legal plates and, of course, no insurance. Such trivialities hadn't gotten in my way before, and I had already ridden around the block a few times without incident. Emboldened by my success, I decided to ride over to my girlfriend's about five blocks away.

The visit itself went well—the ride home not so much. As soon as I mounted the machine parked in front of my girlfriend's house, one of those familiar white, green, and black Chevy Biscayne sedans of the NYPD variety pulled alongside. A request issued from the officer on the passenger side: "Could we please see your license and registration?"

It was decision time. I could either provide my license and then try to talk my way out of the not-so-minor technicality of operating an unregistered and uninsured vehicle on the highway (which, in the state of New York, was a misdemeanor punishable by a one-hundred-dollar fine, a one-year revocation of driving privileges, and maybe even jail time) or do something totally stupid like make a quick getaway.

It's astonishing how stupid a nineteen-year-old can be!

But I had a strategy—to head up "the Hill," as it was known in my neighborhood, where an old nineteenth-century mansion lorded majestically over—actually, a quite forlorn old mansion loomed over its former manor now filled with modest one- and two-family homes of the proletariat. Here I would have the advantage, since the only way around the old mansion was an alleyway far too narrow for the patrol car. It was perfect for a motorcycle getaway! What I hadn't counted on were the large, exposed roots of the ancient beech tree that catapulted me over the handlebars to a painful, bone-jarring landing on brick-hard dirt. Before I could get back on my feet, I was staring up into the faces of two of New York's finest—very, very unhappy faces. To their lasting credit, the men in blue offered me a choice, which, considering what I had just put them through, was rather decent of them: "Asshole, are you going to cooperate with us or not?"

The day of my court appearance finally arrived. The best that I could hope for was that the judge would view the whole incident with a sense of humor and reduce the charges somewhat, but, quite frankly, I wasn't overly optimistic. One look at the dour expression that permanently disfigured his countenance, and I knew that a favorable outcome was unlikely. I was right. After hearing the officers' account and my pitiful apology, the judge issued a stern reprimand from his bench that could easily have been an Old Testament, wages-of-sin-is-death sermon from a Lutheran pulpit of a type to which I'd been often subjected. The inevitable wages of my sin were that the New York State driver's license that I had so cherished for the last three years was gone! Maybe it wasn't quite death…but close.

But in anticipation of this most likely of outcomes, I had already carried out the master plan that would permit me to continue enjoying the freedom to which I had become accustomed. I'm referring, of course, to the freedom of the road, which, although not specifically delineated in the Constitution, was certainly enshrined in the car commercials of the era (I refer you back to "See the USA in Your Chevrolet!"). In my own mind, I had elevated freedom of the road to the same status as those other guaranteed freedoms and the inalienable right to the pursuit of happiness promised in the Declaration of Independence. I vowed that that I would not be denied my rights, not even for a year—a precious year!

A few weeks prior to my court date and while still in possession of a license, I had driven up to Rutland, Vermont, with my friend to tend to the business of maintaining my constitutional rights. I perused the "rooms to let" in the *Rutland Herald* classifieds, and an hour later provided my new landlord with a modest deposit on a room (it was actually quite cozy—too bad I only got to sleep there one night) in return for a written receipt as evidence of my new address. This document I then submitted to a clerk at the Vermont DMV along with my New York driver's license plus some other paperwork. I walked out with a brand-new driver's license as well as a Vermont motorcycle registration and a beautiful (that little piece of tin was truly beautiful!) green license plate with white lettering. Remember, no liability insurance was required for motorcycles in the Green Mountain State. Wasn't life so much simpler back then?

I recall thinking that here was a state that truly cherished freedom. Maybe not to the extent of neighboring New Hampshire, the "Live Free or Die" state, but close. Vermont, after all, had been home to Ethan Allen and the Green Mountain Boys, who had cherished freedom sufficiently enough to take up arms against the British oppressors and, if necessary, die for the cause. From a more practical vantage, my scheme was, of course, well before computers and the Information Age, and changing residency and procuring a new driver's license could be carried out quite simply. In the '60s, the various united states of America were actually not quite so united, and few states had reciprocity agreements with neighbors in matters of motor-vehicle offenses. Let's face it—even if they had had agreements, pursuing out-of-state offenders was difficult, given the state of technology at the time.

Oh, by the way, my buddy and ofttimes riding companion, Paddy? He had already become a bona fide resident of the state of Arkansas, the "Land of Opportunity"!

My Friend, the Motorcycle Jacket

Once you had a bike, you had to have your motorcycle jacket. And it goes without saying that this meant the black-leather type with zippered pockets and zippered sleeves. Problem was, they were terribly expensive. No way could I just walk into a motorcycle shop and purchase such a garment on my perpetually limited budget. So I did the next best thing: I inherited one. A guy from the neighborhood named Bobby Korfmann had recently sold his motorcycle, a single-cylinder, 500-cc BSA Gold Star, to my friend Bobby Bauer. A little while later, Korfmann bumped into me. "Erv, I hear you got a bike?" I proceeded to tell him about it but didn't get too far before he asked, "You need a jacket? I'm not riding anymore, and I got this jacket hanging in my closet...if you want it, it's yours."

Remember, this was back in the early 1960s, and despite the popular image of the greaser types with black motorcycle jackets hanging around the Wurlitzer, no one I ever knew wore the jacket as just a fashion statement. If you wore one, you were riding—that simple.

"How much?"

"Don't worry, it's old...I'll give it to you."

"Bobby, you just made my day."

"Don't get too excited. Wait'll you see it."

So I went over to his house.

"Here, try it on. I know it'll fit, but like I said, it's not exactly new."

And it wasn't. It already had a few road abrasions, visible reminders of its purpose—and over the next few years, I would add a couple more. The liner was frayed in spots but basically intact, and all the zippers functioned. Inside the top chest pocket (it had a lots of pockets) was an interesting pair of eyeglasses: clear glass with sort of a hinged nosepiece that allowed them to conform to your face when riding.

"Keep 'em," said Korfmann. "They're great for riding!" Which they did indeed turn out to be.

So I tried on the jacket. I remember the heft: it weighed more than any jacket I'd ever worn. To this day, I'm always surprised at the relative lightness of newer motorcycle jackets. They just don't make 'em like they used to! It was a good jacket and over the years would serve its purpose.

I remember the time my friend Bobby and I rode upstate New York, him on his BSA (Korfmann's old bike—we called them "Beezahs") and I on my Harley. We were heading north on Taconic State Parkway and were still an hour from our destination, a friend's house in Stone Ridge, when a torrential summer downpour hit us. My jacket, heavy to begin with, got soaked and must have tripled in weight. At one point, we ducked under the canopy of a gas station, where I removed the jacket to squeeze out the rainwater. It looked more like black ink. The guys at the station couldn't stop laughing. Then off we went, back into the rain. My friend was expecting us.

Upon arrival, we quickly ducked into the house to more laughter. I hung the jacket in the garage overnight and the next day out in the sun, but it was quite a while before it dried completely. Another funny thing: while taking a shower, I saw that my entire upper torso, chest, back, and arms had become a deep, blackish blue: my skin had absorbed dye from the leather. I spent extra time scrubbing away to no avail, and as I stepped out with a towel wrapped around me to show my friends—more laughter! It would be days before that dye finally disappeared!

On several occasions, the hide of that motorcycle jacket absorbed gravel and asphalt that would otherwise have permanently marred my epidermis. That is, after all, what it was designed to do, and behind each scuff and scrape is a story. Here's one of them.

Back in 1965, I think it was, I was riding my Harley northbound on Route 9 somewhere near Poughkeepsie, New York. Route 9 is a two-lane highway that runs along the eastern bank of the Hudson River. Stay with it, and you go to Albany and beyond. It was late summer. I had given myself a few days off from work prior to resuming classes at Queens College. The plan was to ride up to Kerhonkson, an old canal town in the foothills of the Catskills where my family owned some property, and do some hiking and fishing.

Country roads (at that time, the Poughkeepsie area and north was still largely farmland) were always my preference over the New York Thruway. They were scenic and—no small matter—toll-free. It was a beautiful morning with a robin's-egg-blue sky and no rain in the forecast: perfect!

I was cruising along at a relaxed fifty to fifty-five miles per hour, reveling in the prospect of a day or two in the woods. The bike was running flawlessly. Having just signed up for our local Pop Warner Football League team, the College Point ACs, I had also resumed weight lifting and was jogging a couple miles every other day. I was bathing in the euphoria of improved fitness.

Leaning my bike through a gentle curve, a lady hanging laundry outside one of those old Federal-style farmhouses momentarily caught my eye. Ahead of me was more pasture and gnarled old trees casting shadows across the road surface. In this most bucolic setting, without warning, a full-grown collie dashed out on to the highway, resulting in whatever $F = M \times A$ adds up to when Harley meets large canine at fifty.

Next thing, I'm screeching down the pavement, pinned back to the side of the bike on a grotesque sleigh ride. The ride lasted sixty feet, maybe seventy feet before coming to an abrupt halt that catapulted me onto the asphalt for another twenty-foot slide absorbed mostly by my jacket. I would have miraculously walked away without a scratch if my left foot hadn't gotten momentarily wedged between the motorcycle and the road. Staggering to my feet, I felt excruciating pain.

Adrenaline had kicked in, however, and hopping on my right foot as I surveyed the scene, I saw a prone Harley, front wheel spinning, and a lifeless collie in a pool of excrement and blood. In the distance was a young lady, clothespins flying out of her apron, running toward me and screaming—the laundress! My first thought was, *If she starts berating me for killing her pooch, I am not going to take it too well.* But soon, I could make out that she was actually yelling over and over, "Are you all right? Are you all right?"

The young housewife appeared astonished that I was actually upright and even alive. Apparently, she had witnessed the whole incident from her yard and was prepared for the worst. "I've been telling my husband for years that something like this would happen. That dog couldn't stand motorcycles. It's not your fault. Are you sure you're OK? I can take you to the hospital!"

Though my ankle was throbbing, I was actually feeling extremely lucky considering what could have happened, so I assumed I was OK and kept telling her, "I'm OK, I'm OK." Ironically, I was now in the position of trying to calm *her* down!

I even got to feeling sorry for the dog—the one that had nearly killed me—and volunteered to drag him off the highway.

Despite growing pain, I still saw no need for medical attention and instead turned to the bike. I took note of a new scratch here and there and a slight dent in my left gas tank apparently caused by the end of the handlebar. The engine started immediately, but there was a serious leak in the fuel line where it entered the carburetor. It would have to be fixed.

The collie's mistress took me to her house and invited me to sit down on the porch to calm down and make sure I would be OK to get back on the bike. I lowered myself onto her wicker couch and propped the injured ankle up to help reduce the discomfort. She offered a glass of whiskey to calm my nerves, which I did not decline. "Bet you're a good patient!" she remarked. I said I didn't have enough experience in that department to know for sure.

Once back on the bike, I slowly idled to a nearby gas station, where the mechanic loaned me his car to get to an auto-parts store to buy a length of metal tubing that he then flanged for me. After the line was installed, the motor ran without a leak. When I told the mechanic what had happened with the collie, he refused payment. *Very Christian of him*, I thought.

Arriving at the family trailer an hour later, I removed the steel-tipped work shoe from my left foot and gingerly took off my sock, all of which caused considerable pain. The ankle had disturbingly doubled in size. Not good!

That night was basically sleepless because of the throbbing, or maybe an hour or two of rest at best. In addition to swelling, the ankle had turned purplish black. Any plans to commune with nature would have to be put on hold.

The next day, I decided to return home and go to the emergency room in Flushing Hospital for x-rays. No fracture, but the doctor said it was one of the worst sprains he had ever seen, and he was an emergency-room veteran who had seen it all.

It would be six weeks before I walked without a limp. Although I missed the opening game of the football season, I forced myself into starting the next game with a tightly taped ankle. We wound up having a good season, finishing six and two, if I remember correctly.

And another scrape had been added to my friend, the motorcycle jacket—another scrape and another story.

Postscript

A year or so after I sold the bike, my old girlfriend and I were reminiscing about the motorcycle days, the misadventures and close calls, and of course, the collie incident came up. That's when she reminded me about the Saint Christopher's medal.

When I first put the bike on the road, Maddy had given me a Saint Christopher's medal, which she asked me to wear when I was riding. She was Catholic, and that patron saint of travelers still stood in good stead with the church. "I'll do better than that," I promised. "I'll mount it somewhere on the bike."

I should say that I had been part of the confirmation class of 1956 at Saint John's Lutheran Church, and Lutherans weren't really big on saints. Oh, we had a few, but not the panoply that Catholics had. Furthermore, at the age of twenty or so, when the medal was given to me, I was already pretty close to being a confirmed agnostic (no, they didn't actually have confirmation classes for agnostics, but you know what I mean. And agnosticism has even fewer saints than Lutherans do) and wasn't particularly keen on encouraging such superstitions. But a promise was still a promise, so the small amulet was promptly fastened to the front of the motorcycle, where it remained till the day it was sold.

Maddy insisted that the reason I had survived my many close calls was because Saint Christopher had been watching over me.

Despite growing skepticism regarding religion, I didn't have it in me to deny her the possibility that she might be right. "Who knows?" I said. "Who really knows?

The Motorcycle Jacket Gets Me into Trouble

The motorcycle jacket did get me into trouble on one occasion. Here's what happened.

It was one of those August heat waves that New York is famous for, those alliterative "hazy, hot, and humid" days. And when it's hot, that hot town is always hotter! But it was a Friday, and getting on the bikes was always a good way to cool off, so my friend Paddy and I decided to pick up our girlfriends, Maddy and Michelle, after work and start the weekend off with a ride over the Triborough Bridge into Manhattan.

Usually we would cross the East River on the Queensborough Bridge (better known as the Fifty-Ninth Street Bridge), which doesn't have a toll. That bridge would forever be immortalized in the Simon and Garfunkel song "The 59th Street Bridge Song (Feelin' Groovy)." Now, most people know that the expression "groovy" became popular during the 1960s counterculture period to express that something was remarkable or notable ("Hey man, that braless chick is really groovy"). The term can also be found in dozens of other songs of the era, such as "Wild Thing" and "Groovy Kind of Love." But unless you ever rode across that bridge on a motorcycle in the 1960s, you really couldn't possibly understand the song's "59th Street Bridge" alias. Back then, you see, the bridge's road surface consisted of metal grating (no pavement!), and your tires would get caught in the grooves, causing your bike to oscillate left and right. My first few crossings on a motorcycle were rather unnerving (quite frankly, they scared the bejesus out of me!), but after a while, I learned that I was never really in danger of losing control even though it always felt that way. It was just a feeling: "feelin' groovy."

I fear I've digressed...back to the story.

So, like I said, we were crossing the Triborough Bridge on our way to uptown Manhattan, and, like I said, it was hot. So hot that many of Gotham's tenants were sitting out on their fire escapes to escape the heat. So hot that I thought I spotted a Hasidic guy walking down First Avenue without his frock coat and black hat, although my Jewish friends insist that such is not possible. We knew it was too hot for motorcycle jackets, at least, so we just wore dungarees and T-shirts. Our jackets were bungeed on to the luggage racks should we need them later that night.

In the shadow of the bridge on Pleasant Avenue and 116th Street in what was then still a small Italian enclave, Our Lady of Mount Carmel Roman Catholic Church was holding its annual feast. We would stop by, maybe get a sausage-and-pepper hero, maybe a zeppole, and just let the rest of the evening unfold on its own.

We parked our bikes on the corner of First Ave and 116th Street. Although reluctant to leave them unattended (thievery of Harleys being quite rampant), we didn't plan to stay long. As a deterrent to theft, we had long ago installed what were called "Louie locks" on the machines. The lock consisted of a pair of sturdy, four-sided steel nuts welded to the frame and the fork through which a huge, hardened-steel padlock could be inserted to secure the bike in a front-wheel-cocked position... primitive but effective.

It being a hot summer night, we didn't really want to wear our motorcycle jackets, but neither was it a good idea to simply leave them bungeed to the bikes. It was Harlem, after all. We decided it was best to wear them open over our T-shirts.

Soon we were strolling among the food and game concessions, breathing in culinary fragrances and observing local color, when above the street noise, I heard a voice behind me: "Only an asshole wears a motorcycle jacket on a night like this!"

Let me stop right here for a moment and list the options at my disposal.

One would be to pretend that I hadn't heard. It was a fairly noisy street scene, after all. But I knew Maddy had heard—the little flinch gave it away. Or I could have acknowledged what was said and just turn to my girlfriend and say, "Let's just ignore that. He probably comes from an underprivileged family where his father beats him and maybe even his mother every night, and this is how he works out his anger!" Or, "I'll bet that fellow behind us is a political-science major conducting a field study on the limits of free speech." Of course, I could have gone to the other extreme: turn around and punch his lights out.

But I did none of those things.

Instead, I chose what I can best describe as a middle-of-the-road strategy. I would give him a chance to apologize or retract his remark, even allow him to deny directing the comment at me and make something up, like tell me he was just saying to his friends how he felt like an asshole for coming out on such a hot night. I would have accepted pretty much anything. He saves face, and so do I—win-win!

So I told my girlfriend to hold on a minute and turned around to discover that the source of the challenge was an Italian-looking guy about my age, hanging out with his cronies. I remember the "greaser"-style haircut (although no one actually used the term "greaser" back then. I think that was after the movie *Grease*—or was it *Hairspray* or some other homage to that era?) I stepped toward him and asked in a conciliatory manner, "Excuse me, but what did you just say?"

This fellow, of course, also had options. These included a simple apology, which somehow seemed unlikely, but still, it was an option. Let's face it, though—just like I didn't want to look bad in front of my girl, he didn't want to back down in front of his goombahs (isn't it remarkable how analytical one can be about these situations given the luxury of many decades of intervening time!).

My antagonist chose to look straight at me (was that a smirk I detected on his face?) and started to repeat what he had said earlier, word for word, actually: "Only an asshole wears a motorcycle jacket…"

And at that instant, I had heard enough and immediately decked him with a left. He kind of looked at me queerly for a second or two before hitting the sidewalk, and then his friends started mouthing off, and now I was in the middle of a hornet's nest.

To myself, I muttered two short words in recognition of the imprudence of my actions, especially on someone else's turf. Fortunately, as luck would have it, some of New York's finest just happened to be nearby and, sensing the impending donnybrook, got between Maddy and me and the troublemakers. It really wasn't necessary for one of the cops to tell me that we needed to get the hell out of there. I actually had come to that conclusion on my own.

I took Maddy's arm and caught up to Paddy and Michelle, who were just ahead of us in the crowd, eyeing zeppoles with no idea what was going on. "Paddy, let's go. We should be getting back to the bikes!" He looked left and right and saw a swarm of angry guys and a cadre of concerned-looking policemen circled around us.

"What the hell is going on?"

"I'll explain later. We gotta go!"

So the police escorted us back to our bikes, trying their best to hold back the menacing crowd that was yelling threats and obscenities in English with the occasional Italian thrown in—which, for a New Yorker, required no translation. I must say, the men in blue did an admirable job.

Once back at the bikes, I was barely able to put my key into the Louie lock. I was shaking so much and just hoping and maybe even praying that the bikes would start on the first kick. They did, and we took off rapidly through the gears up First Avenue.

Speeding back toward Queens on the bridge, I turned to Maddy who was hanging on to me a little tighter than usual. "I hope you weren't offended by the bad language back there!"

"Asshole," was all she said.

Laughter comes at many moments, and at that one, it burst forth from a place hidden deep within—a place from which all that is primal originates.

The South for the First Time

My first trip to the South had been back in 1954, when I was eleven years old. My parents had had friends who had recently retired to Fountain, Florida, a little town in the panhandle north of Panama City (it was actually close to Alabama). In October of that year, my parents took my brother and me out of school for two weeks (Dad didn't have enough seniority at his job for a summer vacation) so the four of us and our beloved mixed breed, Prince, could go down for a visit in Dad's 1951 Packard with its big straight eight and Ultramatic transmission.

The Illchmanns were not blood relatives, but my brother Willie and I addressed them as *Tante* and *Onkel,* as was customary among close German friends. Onkel had served in the German navy with my grandfather ("Opa") during the Great War, and the families remained close. They had wound up immigrating to "Amerika" in the 1920s to escape the economic and political turmoil of the Weimar Republic.

After retirement from Eastman Kodak in the early '50s, Onkel and Tante relocated from Rochester to Florida, where they purchased a hundred acres of pine forest. A small creek bordered by huge live oaks festooned with Spanish moss meandered through the property. To an eleven-year-old fascinated by the outdoors, every walk through their woods with my new Daisy Red Ryder BB Gun was a magical adventure. Right after stepping onto the trail, coveys of bobwhites would burst out of the palmettos. The trail led to a favorite spot on the creek: a small deep, clear pool where we could bathe while fox squirrels frolicked in the canopy above.

During one of my rambles, a small snake decorated with bright yellow, red, and black bands crossed the path. As a budding naturalist, I already knew from field guides that it had to be either a king snake or the highly venomous coral snake but wasn't sure which. After handling the reptile admiringly (and carefully!) and taking photos, I set it free. Upon returning to New York, I found this little ditty in one of my nature books: "Red touch yellow, kill a fellow. Red touch black, friend of Jack." It was only after the film was developed that I realized it had been a close call!

During our stay, I was sent out each morning to the small chicken coop behind the house to gather eggs, a real treat for a boy from Queens, New York! One morning, the chore of procuring dinner was assigned to me. I was provided with an axe and a brief tutorial, and that was the day that the familiar simile, "running around like a chicken with its head chopped off," took on real meaning!

In 1954, there was something else about the South that wasn't easy to overlook: Jim Crow was alive and well. All over Dixie, we saw signs announcing "colored" motels, "colored" bathrooms, colored this and colored that, or Whites Only this and that. On one occasion, we were enjoying a picnic lunch at a roadside rest (my frugal parents made sure we always had picnic lunches when traveling) when an older "colored" man on a mule-drawn wagon came by. My father stood up and waved hello, but the fellow kept going without so much as an acknowledgment. "Jeez, Dad, that guy wasn't very friendly!" I remarked.

To this day, I recall my dad's response. "Maybe he's just not used to a white man saying hello to him."

Although segregation, America's apartheid, would eventually be legislated out of existence, the changing of attitudes remains a work in progress.

The South for the Second Time

My next trip through Dixie took place during the spring of 1960. Six years older and with learner's permits in our wallets, my brother and I would share the driving under Dad's tutelage. The old '51 Packard had been replaced with a 1956 Volkswagen Beetle: thirty-six horsepower and a top speed of sixty-eight miles per hour (allegedly also the "cruising speed," according to the salesman). Dad had gone on an economy kick and decided that the VW's thirty-five miles per gallon would make a greater contribution to the family fortune than a thirsty Packard straight eight. Dad would also stretch the dollar by haggling prices at a half dozen motels each night before settling on a price suspiciously similar to that of the first motel we had stopped at over an hour earlier. "But look at how much nicer the room is," Dad would point out, "and you boys have a nice pool to swim in." (Willie and I also had certain requirements when it came to motels, which could be summed up by, *Does it have a pool?*)

Back to the VW. It surely lacked the spaciousness of the old Packard, and Willie and I had grown considerably over the past six years, so squeezing four adult-size people with luggage required a keen understanding of spatial relationships (and familial ones as well). And let's not forget Prince (seventy pounds?) squeezed in between Mom and whoever's turn it was in the back seat. The sight of four full-size persons and a medium-size dog exiting the little car at picnic areas and gas stations always provided onlookers with something of a clown-car spectacle!

In addition to reliability and parsimonious fuel consumption, the Beetle clearly had novelty going for it, especially as we traveled through the rural United States in the late '50s and early '60s, where VWs hadn't yet achieved popularity. At each gas-station stop, audiences would gather to give Doctor Porsche's diminutive "people's car" the once-over. For most, the sight of suitcases and gas tank under the front hood, where any self-respecting conveyance should have an engine, was a real shocker.

"Well, I'll be darned...would you look at that!" was often heard from a head-scratching observer. Then, opening the rear compartment to reveal a truly miniscule power source without a radiator would really get their attention.

Beetles were not only funny looking, but they provided humor in other ways. Perhaps most memorable was a prank that my buddy Paddy would frequently pull on College Point's VW owners: a feat of strength performed for the amusement of his friends that involved dead-lifting the front end of a parked VW off the pavement and placing its front wheel up on the curb. All of us would then wait around to see the victim's reaction upon returning to his vehicle, resulting in hilarity on our part. There was usually somewhat less of this on the part of the VW owner. It was all a sort of *Candid Camera* without the camera.

During our youth, who would sit in the front passenger seat of the family car was a major bone of contention between Willie and me. I relished it as the better vantage point from which to spot birds and wildlife. My brother liked to get the front mainly because it denied me a chance to sit there. Now that we were older and had learner's permits, Dad had to occupy the front passenger seat when one of us was behind the wheel, so, again, one or the other was always banished to the back seat. On this trip to Florida, my brother and I actually did most of the driving, with Dad's blessing: "Drive as much as you want," he would tell us. "I've been waiting a long time for you guys to take over!"

So we finally arrived at Tante and Onkel's house at the end of three long days (remember the vehicle's *Hochgeschwindigkeit* of sixty-eight miles per hour), and I still recall the mutual delight over the reunion. The first thing I became aware of was that the chicken coop was still there, but no chickens. "The bobcat vas taking zem all," lamented Onkel. "It vasn't verth it anymore! Now ve buy our ekks like ze ozzer people!" He laughed.

After settling in and catching up on the events of our lives (the Illchmanns were especially interested in how Mom's parents were doing), I took a hike back to that little swimming hole beneath the massive oaks. Bobwhites still burst out of the palmettos, and fox squirrels still hung out in the high oaks above the creek. I found that reassuring.

The next day was a Friday, and Willie and I decided to drive up to the little town of Fountain with its nice freshwater lake and swimming area. We wouldn't require an adult licensed driver to accompany us, because the legal driving age in Florida was sixteen, At least, that's how Onkel understood the Florida Vehicle Code, and my brother and I saw no need to question his interpretation.

At the "beach," we struck up a conversation with two sisters. They introduced themselves as Sharon and Becky, and at ages sixteen and eighteen, they epitomized southern womanhood... as if we had any way of formulating such a concept. But they were cute, especially in bathing suits, and friendly. Very friendly, actually. "Why don't y'all come over tonight and hang out with us at our house?" they suggested—an offer we immediately accepted. They got into a '53 Ford, and we then followed them

Escape from the USA

to where a dirt road with eight mailboxes entered the highway. "Drive down this road about a half mile to where it ends. We're the very last house. About seven o'clock?"

We told them we would be there at seven.

When we arrived back at the Illchmanns' house, we shared our evening plans with my mom and Tante, explaining that we had discovered genuine southern hospitality. Onkel, however, had already expressed an interest in test-driving the Beetle and had disappeared with my father, and there was no telling when they would get back. "Don't worry," Tante offered, "if they're not back in time, you take my Plymouth." No ordinary Plymouth, it was a 1957 Fury with the 318-cubic-inch V-8, push-button-drive TorqueFlite transmission and torsion-bar suspension: one of the hottest cars on the road at the time. We thanked her profusely and secretly hoped that Dad and Onkel would not return from their test-drive anytime soon—which they didn't.

Willie slid behind the wheel of the Fury and entered the highway, rather conservatively, I thought. A half minute later, he punched the gas pedal until the speedometer reached a hundred, and in no time, we were ten miles up the highway at the dirt road with the eight mailboxes. Sure enough, at the end of the road stood house number eight. It was one of those basic frame structures perched off the ground on concrete footings so common throughout the rural South. A galvanized tin roof with that typical rusty patina still appeared capable of keeping the rain at bay. The front of the house was covered by clapboard, and only the front. The rest was covered in tar paper, and one couldn't be blamed for thinking that the siding project wasn't going to be finished any time soon. Somewhere in the past, a burst of creativity had culminated in a row of truck-tire planters separating a red dirt driveway from a nonexistent lawn—or at least I thought they were planters. They held no actual plants, so it was hard to say for sure. The '53 Ford was parked outside, and not ten feet away was an identical model, tireless and perched on metal milk crates.

Just as Willie parked the Fury, the girls came out to invite us in for formal introductions. The interior ambience could best be described as mid-twentieth-century tobacco road. Tired linoleum covered the floors, and front pages from the *Panama City News Herald Sunday Magazine* section were tacked to the walls, giving the place the look of a hillbilly art gallery: there was Elvis, smiling in his army uniform; Jackie Kennedy smiling in her pillbox hat; and Ben Cartwright smiling in his cowboy hat.

We got to meet Ma, Daddy, little sister Rachel (a skinny twelve-year-old), one-year-old Mary Lou (who turned out to be Becky's daughter from a short-lived marriage), and Granddaddy Earl wearing a pair of those one-piece denim bib overalls over a white T-shirt...country indeed.

Bowls of potato chips and pretzels were on the kitchen table, and soon we were drinking cans of Busch Beer. Daddy and Granddaddy Earl had lots of questions about what it was like to live up in New York—and did we ever get mugged by Negroes (pronounced *niggras*). But they did have a TV in what served as the living room—black and white, of course, maybe with a sixteen-inch screen—on a movable cart. The *Gillette Cavalcade of Sports* was on, featuring a world middleweight-boxing championship bout between Gene Fullmer (the "Mormon Mauler") from Utah and Joey Giardello (actual name: Carmine Orlando Tilelli) who had grown up in the Bedford-Stuyvesant section of Brooklyn, New York. The fight was being televised from Bozeman, Montana—actually the first and only championship fight held in that town.

As the contestants were being introduced, Earl commented, "Good to see two white boys in the ring...ain't right that a white boy should be boxing a Niggra!" Regardless, it turned out to be one of the dirtiest fights in the history of boxing, with both men head butting and brawling their way through fifteen mean and ugly rounds. Perhaps the best thing that can be said about the contest was that it ended in a controversial draw, since declaring either fighter a victor would have made for a greater travesty.

Controversy would follow Giardello well into his retirement years. In 1999, Universal Studios released *The Hurricane*, directed by Norman Jewison, the story of middleweight boxer Rubin "Hurricane" Carter. Carter had been convicted in 1966 of committing a triple homicide in a Paterson, New Jersey, tavern. His conviction became a cause célèbre among those who thought racial bias had been behind the verdict. The film itself played fast and loose with facts (yes, folks, Hollywood sometimes does that sort of thing), including its depiction of the Carter-Giardello championship bout as a blatantly racist decision. In the movie, the Giardello character, despite receiving a severe beating at the hands of Hurricane Carter (played by Denzel Washington), is awarded a unanimous decision by the judges.

The only problem was that Joey Giardello, now almost seventy years old, was not only very much alive but also very much offended by how he had been portrayed on the big screen. So upset, in fact, that he did what any self-respecting Brooklynite would do: he sued the bastards. During court testimony, most boxing analysts had thought Carter lost. In an interview, even Carter himself said he had lost. Giardello was awarded damages for the unfair depiction of the fight, and Jewison agreed to make a statement on the film's DVD version that "Giardello no doubt was a great fighter."

But I think I may just have digressed again. There's a motorcycle adventure to talk about.

Escapism

I explored the concept of wanderlust previously as an explanation for our desire to travel, but what about *escape* as an equally potent force? Look at the glossy advertisements in travel magazines: "Come join us on a Caribbean escape!"

My friend Paddy was dealing with a failed marriage, and I was struggling to find a way to end a relationship that was going nowhere (but on some irrational level, I still held out hope for it). Insecurity was the modus operandi dictating both of our lives. Paddy couldn't or wouldn't share anything with me about his failed marriage even though I had stood as a witness during his civil ceremony, painfully aware that the union was doomed: an ill-fated venture in which I had been complicit.

We both imagined, I suppose, that in some weird way, the road trip would have therapeutic powers. My friend would be able to exorcise the failed-marriage demons, and a prolonged absence from my on-again, off-again girlfriend would provide me with the courage to move on: the time-honored, "we need some time apart" strategy. I was convinced that the journey would put a merciful end to a flawed relationship. (And that end did eventually come, but not because of the road trip.)

So, off we would go. Free to plot our own itinerary, we chose state and county highways over the rapidly expanding Dwight D. Eisenhower Interstate Highway system. On these roads we could enjoy the simple pleasure of riding into a small town for

Escape from the USA

the first time. Let's hear it for life's simple pleasures! Those cozy, blue highways on our torn and frayed (but always free) road maps that traced our nation's trails of westward migration and followed the land's contours, unlike the Interstate system that was engineered to dominate and conquer with an agenda of commercial and military efficiency. Small two-lane roads that embraced towns and villages, inviting travelers to pause and find a root-beer float at a mom-and-pop soda fountain or take comfort in the elm-tree shade of Main Street while Main Street, USA, still existed. Maybe Interstates were becoming the arteries and veins of our sprawling nation, but the blue highways were still its capillaries where life really took place. Walt Whitman's 1856 "Song of the Open Road" never comes to mind on the Interstate!

> Afoot and light-hearted, I take to the open road,
> Healthy, free, the world before me,
> The long brown path before me, leading wherever I choose.
>
> Henceforth I ask not good-fortune—I myself am good fortune;
> Henceforth I whimper no more, postpone no more, need nothing,
> Strong and content, I travel the open road.

Erwin "Erv" Krause

Krause Haus in College Point

Escape from the USA

Poppenhusen Monument

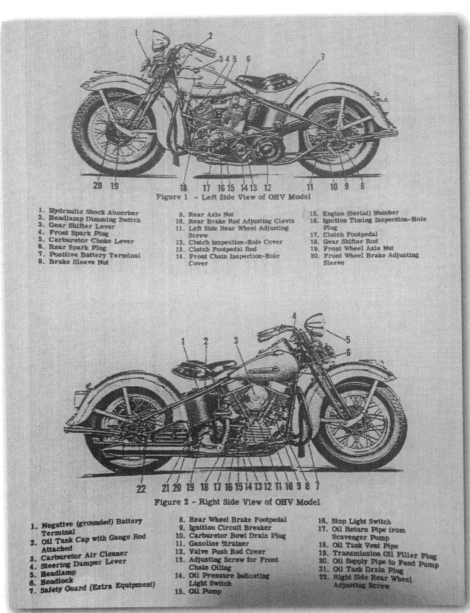

Harley diagrams

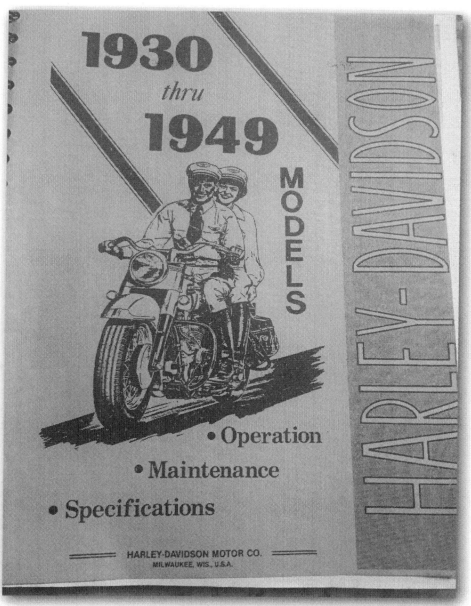

Harley Manual

Erwin "Erv" Krause

Harley Manual 2

Escape from the USA

My ol' forty nine

Erwin "Erv" Krause

My old Harley

Escape from the USA

The cost of rebuilding

Roadside adjustments

Escape from the USA

Clowning around

Paddy and Erv in Texas

All Come to Look for America (The Times, They Are A-Changin')

"Be careful down there," we would often be warned, "they're still fighting the Civil War, you know!" And in many ways, a journey through the South in the '50s and '60s, at least to me, was like traveling through a foreign country—one still smoldering in the aftermath of military defeat. As a Northerner—and I know this sounds a bit crazy—I sometimes harbored that awkward feeling that I imagine soldiers might experience when marching through conquered territory.

Early in the nineteenth century, the South had increasingly alienated itself from the North by tenaciously holding on to "that peculiar institution" of slavery. To defend its right to maintain the bondage of a race of people, the Confederacy initiated a ruinous conflict costing thousands of lives. The humiliation of military defeat was followed by the added ignominy of Reconstruction, although the "redemption" period and the adoption of Jim Crow laws may have offered some recompense.

But this was 1966, and the Civil War had ended a hundred years ago. Right? But in the view of many southerners, a new onslaught of "Northern aggression" had been launched: the civil-rights movement. Yankee agitators, mostly Jews, it seemed, along with uppity Negroes and nigger lovers, began poisoning Dixie with ideas of equality and integration and a host of other offensive "isms": agnosticism and its blasphemous relative,

Darwinism, and without a doubt sneaking in some atheism, socialism, and communism. Too bad for those two Jew boys Goodman and Schwerner and their nigger companion, Chaney, but they had it coming!

This was the South we would be traveling through: the wounds of old "Northern aggression" were just starting to heal, and now they were being forced to defend themselves against a new assault on their way of life. Although the last surviving Civil War veterans had passed away in the 1950s while I was still a youth, there were still thousands of southerners who had sat at the knees of daddies and granddaddies who had taken a rebel stand and had heard their stories: stories of civilian hunger and other deprivations inflicted upon them by Grant and the heinous William Tecumseh Sherman.

A humorous aside about the attitude that many northerners harbored toward southerners came from a well-meaning acquaintance advising me about travel in the South: "And one more thing I should warn you about: be careful of any guy with the middle name Lee—like Bobby Lee this or Sonny Lee that."

"Why's that?" I had asked.

"Go to the post office and check out those Most Wanted posters on the wall."

During the '50s and into the '60s, US post offices, as a public service to the community, would display wanted posters with pictures and descriptions of notorious felons on the loose. As a kid, I remember being fascinated with these and would flip through them while waiting for my mother and father to finish their postal business. But was there anything at all to this fellow's assertion about the middle name Lee? At the next opportunity, I checked the posted names of the wanted scoundrels, and to my astonishment, many of them—certainly more than you might expect—did indeed have this moniker as their second name!

Escape from the USA

Perhaps not the most scientific inquiry, but I made a note to myself nonetheless.

Mindful that I was, after all, a Yankee, and there was no way of disguising this (remember those Vermont plates on my bike), I would be cautious. "Just behave yourself," I had been advised. "Don't give them any excuse." Conquered people are often sullen and can be vengeful toward the occupation.

I've already alluded to the assassination of the three civil-rights workers in Mississippi back in June of 1964. One of them, Andrew Goodman, had been a classmate of mine at Queens College (we called him Andy). Queens was a large, almost entirely commuter school, part of the City University of New York (CUNY) system. Its students were mostly Jewish during the 1960s, though I attended because it was free and had a fine academic reputation. I made some good friends at Queens, and (sorry for the cliché) two of my best friends were Jewish, although much of my social life still centered around College Point friends.

Andy was not a "friend" in any real sense, but we had been enrolled in a required phys-ed class together and got to know each other. There always seemed to be a calm, gentle aura surrounding him and a degree of maturity or perhaps sophistication beyond the norm. I respected those qualities, but at the same time, they made Andy inaccessible to a blue-collar, rough-around-the-edges sort like myself. In retrospect, that inaccessibility was of my own doing, a reflection of my own insecurity.

Andy had attended an elite private school, while I was the product of Flushing High School, not exactly a paragon of academic excellence. He had come from a family of educated, progressive left-wingers; my father, on the other hand, voted Republican, and my mom simply followed Dad's instructions upon entering the voting booth. Andy had also devoted himself to progressive causes, but I was unaware of the extent of his commitment until I heard the news that he and two other civil-rights workers had gone missing while working with the Congress of Racial Equality (CORE) on a Negro voter-registration project in Mississippi. It was summer recess, and I was working at my summer job as a construction-crew laborer in Brooklyn when I first heard the report. This is a thing you never forget. The bodies were found buried in a levee two months later. Most of the KKK people responsible got away with murder.

I shouldn't go any further without a disclaimer: there is no comparison between Andrew Goodman's trip to the South and my own. Andy was on a mission, a righteous mission, into the Deep South—the heart of darkness that was Mississippi in the '60s. My friend and I would only be traveling through the South on the way to some vague destination, hopefully Mexico. Whereas Andy and his companions had aspired to change the South, I simply wanted to see it for what it was, warts and all. Andy had been a missionary with zeal. I was more a casual anthropologist seeking adventure and a change of scenery. Maybe eat some barbecue or partake in catfish and hush puppies, foods you couldn't get in New York back then. Maybe visit one of those snake-handling church services or participate in a Pentecostal service and speak in tongues right along with them folks. Meet some southern lasses and listen to their music. Hike through cypress swamps and see gators and snakes. Just have fun and be a part of the growing number of young people celebrated later in song by Simon and Garfunkel who "all come to look for America" and do it in our own crazy way.

Though outraged and saddened by Andrew's murder, I can't say I was surprised. Ultimately, what killed Andy was his goodness, and he was a truly "good man," as his surname implies. That irony was not lost on me. Possessing not a trace of malice himself, he was incapable of seeing evil in others, making him unlikely to understand the peril he was in until the hammer was cocked. Or so I was convinced when I learned of his murder.

But what I did not realize back then and only came to understand later in my life was that this gentle, idealistic New York Jew was also one of the most courageous men I ever knew.

In 1972, singer-songwriter Don McLean rose to the top of the charts with an eight-and-a-half-minute-long ballad: "American Pie." The song was filled with arcane references, mythology, and symbolism the meanings of which were discussed and debated for the months of the song's popularity and well thereafter. To this day, there is still no total agreement on their meaning, but at least today, you can Google the song and get some insights, something we could not do back then. But the meaning of the song's refrain in which them good ol' boys were drinkin' whiskey and rye (and I don't care what anyone, even McLean himself, says), I am certain of.

The Great American Road Trip

We would soon embark upon that quintessential American experience, the road trip—but on motorcycles, not in a car. And, of course, there's a world of difference. Consider conversation. You may exchange words with a travel companion while riding, but such verbal interaction hardly qualifies as conversation. On a bike, it's staccato bursts concerning matters of immediacy: "Let's eat...I'm hungry...stop at the next gas station...engine's making a funny noise...see that babe in her DeSoto...skirt was hitched up nice and high!" Not for big talkers, the motorcycle trip. If you have a lot to say, wait'll you pull over to catch up!

And, let's face it, riding is inherently more dangerous than driving a car—no need to review the basic laws of physics. Hit a woodchuck while driving a Buick at sixty-five, and you might say, "What was that?" Hit one on a bike, and it could be a life changer.

Well-meaning folks advised, "Drive defensively!" But I always took it a step further: "Ride paranoid...they're out to get you!" And my experiences on the highway continue to inform me that this was no exaggeration. After all, even paranoids are right some of the time! Of course, if you always ride in such a state of mind, it will suck out the joy, so you need to remove that negative thinking from your consciousness...but not entirely! How about we just say that for the motorcyclist, there's always a heightened sense of awareness. Some call it "situational awareness" others might say exteme "mindfulness." No matter. It's a skill you need to learn; ask the survivors.

Because you ride under the sky and just slightly above the earth, senses are heightened anyway, for better or for worse. Very aromatic, the motorcycle ride: there's the burning-rubber smell of skunk; pine forests; alfalfa fields; dairy farms; the oil refineries of Bayonne, New Jersey; and the pulp mills of New Brunswick, Georgia. Other senses too. You feel the coolness of valleys. The simple sensation of air in your face and hair makes motion itself more real. The sun is an almost constant companion: warming your back when heading west in the cool of the morning and then taking its toll on your face in the afternoon… which is why they made Noxzema.

And then there's those "earworms"! Maybe it's the pulsating of the engine or the rhythmic beat of expansion joints on concrete highways. Or maybe it's simply because you're alone with your thoughts on a motorcycle. But riders are more susceptible to that "I can't get that song out of my head" phenomena. Annoying or soothing?

But there's an eight-hundred-pound gorilla in the room: *danger*! I've alluded to it already, but let's address the subject more directly. Back when I was riding, I would often hear from nonriders—those wary, cautious, or maybe just sane individuals—"Oh, you got yourself a murdercycle?" I was even told by one former rider that each time you got off your motorcycle at the end of the day, you had "cheated death." Although I wouldn't take it that far, there is no denying the element of danger. Not only is it real, it's actually intertwined with the thrill and the joy of riding. The statistical reality is quite simple: every time you get on that motorcycle, there actually is greater risk than behind the wheel of a car. In fact, it was during our journey south that an event occurred that is now part of pop-culture legend (some say mythology): Bob Dylan's accident while riding his Triumph on a country road near Woodstock, New York. Some news reports had him near death with broken neck vertebrae. Others, to this day, claim the accident was faked or perhaps exaggerated to enable him to escape the mounting pressure of career demands.

I was reminded of this danger factor in my late thirties (my children Eric and Paul were still in elementary school), when I did the prudent thing and applied for term life insurance. The company required a physical, a health history, and the answers to questions about lifestyle: cigarette smoker? (*No!*) Drink alcohol? ("In moderation," of course…which, considering I was from College Point, was a truthful response.) How about recreational activities: check the yes or no box. Scuba diving? Mountain climbing? Skydiving? And there it was: *Motorcycle riding?*

I asked the insurance rep what would happen if I checked the yes box for motorcycling. Would I be declined?

"Not necessarily, but there would be a premium adjustment."

"Adjustment…upward, you mean?" (Stupid question, I know.)

"Of course!"

I knew that, but I had never really given the issue of mortality much thought until children entered my life.

The Departure (The South for the Third Time)

A few days before our scheduled departure, Paddy and I put together a list of things to bring along on the trip. Here's what it looked like:

Motorcycle Stuff

assorted tools
spare spark plugs
spool of electric wire
flat-tire repair kit

two quarts Harley oil
small socket wrench kit
guide to Harley-Davidson dealers
one can STP oil additive

Camping and Misc

matches and cigarette lighter
small grill for cooking
US Army surplus mess kit
air mattress
US Army surplus sleeping bag
canteen

ground cloth (tarp)
hunting knife
Emerson transistor radio (red)

Erwin "Erv" Krause

PERSONAL

dress shirt	bathing suit	road maps
T-shirts and underwear	towel	
extra pair dungarees	ditty bag	
socks	English Leather a/s	
sneakers	address book for postcards	

We had road maps to help us navigate the eastern United States; along the way, we could pick up any other maps we needed at gas stations—they were free!

There was another list as well—of things we thought at first would be cool to have, but upon reconsideration, we left behind, such as an extra dress shirt and pants, cooking gear for the open campfire, and a tent. What we actually stuffed into our saddlebags and bungeed onto the luggage racks was enough. The night before our departure, I painted

??? or Bust!!

on each pannier to announce the flexibility of our itinerary—and to remind ourselves as well.

Everything was packed the day before; we didn't want to be fumbling around in the dark the morning of. I would meet Paddy at his house (about eight blocks away) at 4:00 a.m. We hoped for an early start to beat the rush-hour traffic. Then we would head south on the Jersey Turnpike and make decisions along the way.

Mom and Dad woke up with me at about two thirty in the morning so we could have breakfast together, which my mother insisted upon—although I didn't really need any convincing. "You can't go anywhere without a good breakfast," she would always say. This was more philosophical advice than nutritional.

Escape from the USA

Over breakfast, I went over our travel plans, which, as vague as they were, couldn't really be called an itinerary.

"Try to stay in touch...you can reverse the charges," Mom requested. I assured her I would.

Mom made pancakes (*Pfannkuchen*), eggs, and bacon. I ate it all—of course I did. I always ate it all. She also packed lunch for me and my friend. Food is love and, in my mother's opinion, its highest expression—so why not pack along some love in the lunch box?

Dad got in some final advice. "Remember what I always told you. You don't want to be out on the road at closing time! Make sure you stay in the right lane except to pass...I shouldn't have to tell you that by now. And if you get tired, pull over and take a nap. Don't take chances. We want you back in one piece. And don't give southern cops any excuse...they'll pull you over in Georgia faster than you can say William Tecumseh Sherman!" He then repeated the oft-told story about a friend of a friend who had spent the night in a Georgia jail for walking across the street against a red light—jaywalking! He also rattled off a few notorious speed-trap towns and how they targeted out-of-state plates.

And what of the real thought that my parents had concerning their son's gallivanting around the country with his roguish companion on their Harley-Davidsons? We did have something of a track record of mischievous behavior, after all. But if they had misgivings, as they surely must have, these feelings were not shared with me. They understood the importance of a proper bon voyage, and whatever doubts and trepidations they had were kept to themselves.

I walked out into the early-morning darkness with my dad, where the melodic chorus of robins was already underway from the maples and sycamores that lined our street. Under orders never to start the bike in the garage, I wheeled it out into the driveway, gave it a few kicks, and turned on the ignition switch. It started on the next kick…a good omen. I was always looking for omens. Rumbling to the corner, I turned around for one last wave good-bye and headed up Fourteenth Avenue. In an hour, Paddy and I would be "counting the cars on the New Jersey Turnpike" on our way to look for America.

Barnyard Humor and Hospitality

Riding blissfully through Appalachia's serpentine roads, throwing the big bikes into curves and powering through and downshifting before the next bend—and repeating and repeating and repeating—if you don't have a smile on your face, then perhaps you should consider another form of transportation (or maybe just check to see if you have a pulse). Could I be forgiven for thinking that in those moments, I had discovered the key to happiness?

One ascent with lovely banked turns gave us a chance to really lay the bikes down, footpads scraping pavement and throwing off sparks like a Roman candle. Laughing so hard it was foolhardy to continue, we got off the bikes to get the hysterics out of our systems. Once back on our bikes, while motivatin' over a hill, we came face-to-face with an ominous sky, so dark you'd think evening had arrived. But that couldn't be in the afternoon at three. Then came a telltale drop in temperature just as the sky ahead lit up, followed a few seconds later with rolling thunder. Fifteen seconds later, another flash of lightning followed even more closely by a serious thunderclap. We were heading into the storm!

Racing downhill through patches of woodlots and pastures bordered by ancient stone walls, we saw a farmhouse and barn ahead just as fat globules of rain appeared. The barn with its faded Mail Pouch Tobacco advertisement would be our refuge.

The interior lights illuminated the farmhouse against the dark, sodden sky. A fifties-era Ford pickup sat in the shale driveway. The place was clearly occupied. Had we been seen pulling into their barn? Perhaps not, since thirty or so yards separated the two buildings. But perhaps we had been...then what? Would we be welcomed or maybe tolerated, or perhaps simply ignored? Then again, there was that possibility of a hostile reception; you never knew. We were now in the South.

For the moment, at least, we were secure in a shelter, so we shook the water off our jackets and climbed up the ladder into the hayloft. We had to raise our voices, so loud was the deluge hammering down on the tin roof. The feeling of good fortune at pulling into the barn just before the skies opened was tempered by uncertainty; we had entered private property without permission, after all. But what about the tradition of giving mariners safe haven in the event of a storm? Didn't it apply to travelers on terra firma?

These were things we had time to consider while reclining in the hay and entertaining ourselves with jokes in "the farmer's daughter" genre. "So the third guy says, 'My name is Chuck.' And the farmer takes his shotgun and shoots him!" In the midst of our laughter, the farmhouse door slammed open, and we saw the farmer himself bundled up in a yellow slicker and walking rapidly in our direction.

Thinking the worst, I looked toward Paddy and mouthed, "Oh, shit." On the plus side, this farmer wasn't carrying a shotgun. He was tall and sort of gangly, and in no time, he stood in the barn looking up at the two of us.

It was always best to be proactive: "Sorry we didn't ask permission, but with the heavy rain…"

Escape from the USA

He stopped me. "No need to apologize. The missus and me saw you boys comin' down the road and the weather you was up against. You're welcome to stay right where you are... just wonderin' whether you'd like some cookies. Sharon"—he said "Sharon" like we already knew her—"just baked them this morning. And some milk?"

Paddy and I looked at each other for a second. I was actually feeling a little guilty about conjuring up those negative stereotypes about southerners and maybe even the farmer's-daughter jokes we'd been sharing. We thanked him for his kind offer, but seeing as how it was still pouring, we didn't want him going back and forth in the rain on our account. We assured him we would be on our way as soon as it let up.

The noise of the rain made discussion challenging, but we managed to tell him where we were from and a little bit about our travel plans. He told us a little about himself and the dairy farm. A nicer guy you couldn't ask for. "Well, you boys stay as long as you need," he said, and then he jogged back to the house, where we could now see his wife standing out on the front porch, arms folded, probably hoping everything was all right out there.

When he was well out of hearing range, Paddy pretended to call out to him, "Hey, by the way—no cookies, thanks, but you wouldn't happen to have any daughters, would you?"

Outside Agitators

That's what they were called, those folks from up North who came down to try to change the "southern way of life," which, by the way, had worked perfectly well for all these years when everyone knew their place. The John Birch Society–sponsored "Impeach Earl Warren" billboards began to appear with ominous regularity. Did it occur to us that we might be perceived as "outside agitators" during our peregrinations through Dixie?

Unlikely. Outside agitators were mostly Jews or "faggots" and didn't ride Harleys. If folks down here started talking politics, we would remain neutral like Switzerland. This was our strategy. And it would be tested shortly after we arrived in the South.

It happened while we traversed the lovely state of Georgia, proud of its famous peaches, including that most ornery of all peaches: Ty Cobb. Ray Charles had already let the world know how much the state was on his mind, but Gladys Knight and the Pips hadn't yet boarded their midnight train. We, on the other hand, had occasion to tarry at a Harley-Davidson dealership well north of Atlanta, lured in by a pair of gleaming new Electra-Glides sitting in a bare-bones showroom. And while we were at it, why not some routine maintenance (adjust chain tensions, check spark plugs, and change the oil)? We wouldn't need the services of a mechanic, since we always did our own work—just some quarts of "official" Harley-Davidson oil, and, if you would be so kind, a place to perform the tasks, please? With southern hospitality to spare, the manager directed us out back to a shaded area beneath a huge live oak.

The soil beneath the tree (which, judging by its girth, must have dated back to James Oglethorpe) was besotted with decades of waste oil and gasoline. The massive specimen appeared to be thriving on its fossil-fuel fertilizer regimen. Not far from our workstation was a fenced-in area with discarded Harley parts: engines, pistons, trannies, frames…you name it. A motorcycle necropolis or a treasure trove, depending upon point of view, from which a creative gearhead could glean enough parts to assemble a small fleet of viable machines. Or at least such a thing seemed possible in my imagination. Paddy, in fact, had already glommed a primary chain and stuck it in one of his saddlebags. Hey, you never know!

While we worked on the bikes, one of the mechanics joined us during his lunch break and began to shoot the breeze. "Where y'all from?"

"New York." (Probably sounded like "New Yawk" to him.)

"Ever get mugged?" (You wouldn't believe how often we were asked that question!)

"No. It's not as bad as people think."

And so on.

After a couple of minutes of chatting, he pulled out of his pocket what appeared to be a folded-up newspaper, which it actually was…of sorts.

"Ever see this paper? It tells the truth…the *real* truth, about what's goin' on in this country!"

It was called *The Flaming Cross*, the official news organ of the Ku Klux Klan (KKK), which was still alive and well in many parts of the Deep South. Our Harley-mechanic acquaintance handed the paper to us. "Y'all need to know the truth about what's goin' on down here…things they won't tell you in those [Jew] newspapers in New York."

Erwin "Erv" Krause

He directed our attention to a prominent black-and-white photo of a racially mixed gathering of young people, male and female, assembled on the marble steps of a southern courthouse. Signs and placards lying about made it obvious that they were civil-rights workers, probably a small contingent of so-called freedom riders who were traveling to the South in the '60s to fight for voter registration and school integration: the infamous "outside agitators." The mechanic went on, "You gotta read what it say in order for y'all to understand what them people are doin'. They won't tell y'all about any of this up North!"

The caption beneath the photo read something like, "This photograph shows an example of the disgusting acts of depravity taking place in public and in broad daylight by these outside agitators." I had to look closely to figure out exactly what "act of depravity" was being depicted, and then I discovered it. Among the young people on the courthouse steps was a tall Negro standing with his back to the camera and, partially blocked from view, a seated young, blond white girl. These two were most likely just talking. But in the perception of a southern bigot, there was clearly sex going on here. How could it be otherwise?

I started to respond with an explanation involving camera angle and how things aren't always what they seem when it comes to photographs, but I stopped myself. "Stay neutral like Switzerland," a little voice reminded me. Nothing good could come from continuing the discussion. Two guys from up North, now in enemy territory and vulnerable, on motorcycles. I changed the subject. "What do you do with the waste oil?"

"Jes' leave the awl on the sawl...keeps the dust down."

I wanted to address the issue of environmental harm caused by improper disposal of the oil, but for the second time in under a minute, I reminded myself that discretion is the better part of valor!

The Mystique of Grits

Roadside rests are not campgrounds, but they are places where a man can lay his head, and if you like your mattress firm, nothing beats the top of a picnic bench as long as you're not a restless sleeper. Paddy and I discovered this and the creative (we thought so, anyway) idea of inflating air mattresses with motorcycle exhaust. Sure beat huffing and puffing and that light-headed feeling you get from using your lungs (we forgot to pack the bellows). Rest stops usually also had a running water and a sink, or at least one of those old cast-iron water pumps with a long handle, which is all you needed for a "French shower" (yes, I know, that's a politically incorrect remark). That and a splash of English Leather.

Shortly after dawn, we were back on the Gulf Coast Highway, motoring in solitude through beautiful marshland. The air was still somewhat cool, but our jackets blocked the chill and absorbed the heat of the southern sun at our backs, making the experience sublime.

The serenity of the moment was suddenly interrupted by a huge wild hog that bolted out of the marsh grass and dashed in front of me. I braked, of course. There could be others, but missing him was more a matter of luck than anything else. It was a close call! I had seen them on previous visits to the South, but the sight of the huge beast with its ridiculously skinny underpinnings made me laugh. Or was it the nervous laughter of relief that comes from knowing that you just narrowly avoided catastrophe? I had already had that encounter with the collie and a few close calls with deer. Along with the laughter came trembling.

I slowed down to let Paddy catch up. "Did you see that?" I bellowed.

"Yeah, I did. Why didn't you hit him? I'm in the mood for some bacon!"

More laughter.

A half hour later, we were in fact lured into a small café by a "Bacon…Eggs…Grits Special…99 Cents!" sign, not to mention a parking lot filled with eighteen-wheelers; truck drivers know all the good places to eat! In the parking lot, some boys were chatting and enjoying a smoke. They asked the usual "Where y'all from? Where y'all goin'?" questions.

Once we were inside the café's murmuring ambience, heads turned toward us in curiosity. An apron-clad waitress with a bodacious Tammy Wynette beehive placed menus in front of us. "Coffees?"

"Two, please."

Paddy excused himself, leaving his instructions—"When she comes back, I'll take the special but toasted white instead of grits"—before trotting straight to the bathroom with a clear sense of urgency. Seconds later, Tammy returned with two coffees and, in twangy Americana, asked, "Now, how can I help you, honey?"

I loved the flirtatious way that southern waitresses took orders, and the first few times, I fantasized that some sort of romance had instantly blossomed. But that fantasy withered on the vine when it became apparent that every other guy was also "hon." Still, those southern girls, "with the way they talk(ed), they knock(ed) me out when I (was) down there."

"Two specials. But could you give my buddy here"—I pointed to the empty stool—"toasted white instead of grits?"

"That won't be no problem, hon." I also had a weakness for girls who used double negatives...so different from the girls at Queens College.

Paddy returned to his stool just in time for the breakfast specials. I looked at him with some concern. "You look perturbed... anything wrong?"

"What'd we eat yesterday?"

"We stopped in that catfish-and-hush-puppy place, remember?"

"Oh, yeah...where the bartender gave us that alligator mating-call demonstration."

"You always amaze me...how can you possibly remember stuff like that?" It was apparent that he was blaming me for his stomach issues because of my penchant for trying southern cuisine wherever possible. "You and your 'When in Rome, when in Rome.' You know what, tonight, let's pretend we're *actually* in Rome and just get some pizza!" Then he looked at my plate. "Since when you like grits?" (He enunciated "grits" with clear disdain.)

"Since we crossed the Mason-Dixon line. When in Rome..."

"What are they, anyway?"

"You mean where they come from? I have no idea...maybe a grit plant."

"They have no taste, man! Might as well put salt and pepper on a napkin and eat it!"

"I admit, they require a little seasoning..."

"A little?"

"Season to taste."

The Get-Out-of-Jail-Free Card

One hot, sultry evening, we pulled into the little fishing village of Cameron, Louisiana, county seat of the eponymously named Cameron Parish.

Some background information: back on June 27, 1957 (nine years prior to our arrival), the town had experienced its defining moment when Hurricane Audrey roared in from the Gulf of Mexico in the dark hours of early morning with a ferocity that no one was prepared for. With winds as high as 180 miles per hour and a storm surge of twelve feet (some say twenty!), the storm killed over four hundred people and destroyed or damaged between 60 and 80 percent of the community's homes and businesses. So devastating was it that the weather service officially retired the name Audrey as an identifier of Atlantic hurricanes.

Following the cataclysm, local history would henceforth always be discussed in the context of before- and after-Audrey periods. Conversations would often begin with "Before Audrey" or "Since Audrey." One doesn't hear that reference point used much in the twenty-first century, but ironically, it took the arrival of another killer storm, Rita, in 2005 (nearly fifty years later) to change that.

I should say that we knew absolutely nothing of Cameron's history before we arrived. We actually got educated somewhere between the hours of 9:00 and 11:00 p.m. while attending a

Escape from the USA

lecture conducted by a local historian who also happened to be the bartender at the tavern we'd stepped into upon arrival in town. The main purpose of our visit to that bar was to inquire about budget-priced lodging. Every two or three days, we had gotten into the routine of giving up our cozy sleeping bags for a hot shower and a roof over our heads, and we had discovered that bartenders provided excellent guidance in that department. This was perhaps a legacy of days past, when taverns and inns were one and the same along stage-coach routes.

Something did strike us as a bit odd when we'd first rolled into town. In the 1960s, most rural southern towns consisted of a collection of older homes of the unpainted-clapboard, tin-roof variety. Downtown areas of larger towns featured old limestone-block or brick structures, substantial buildings often with the year of construction or a merchant's name proudly displayed on the front gable. None of this architecture in Cameron. Most structures here were quite new and of prefab variety. Even the piers and wharves to which the local shrimp fleet was tied appeared to be of recent construction.

After Paddy and I ordered some cold beers, I broke the ice with something like, "Nice little town you got here…"

"Shoulda been here about nine years ago," replied the bartender, who then proceeded to tell us all about the catastrophe of 1957. John was his name. He gave his last name also, which I have long since forgotten, except for the fact that it was French sounding, like most of the names there.

Even with the storm nine years behind, it couldn't have been easy for John to talk about it, but he certainly shared a great deal with us that night. He was joined at times by one of the four or five patrons present, who nodded frequently and contributed some of their own memories. Although none of the men there had lost any immediate family (probably making it easier for them to share their stories), all had lost friends or distant relatives. Like in so many rural communities, many Cameron residents were related to one another as evidenced by how often certain surnames appeared on mailboxes, businesses, and cemetery plots: Broussard, Erbelding, Henry, Fontenot, and Richard.

The men at the bar, clearly regulars, told stories of houses swept off foundations and winding up miles away, often with survivors clinging to the rafters; of people who had survived by hanging on to the branches of large oak trees while clutching their children; of water moccasins and other swamp creatures slithering among survivors; of graveyard caskets floating away. A great deal was made of how the storm had caught everyone by surprise. It had been supposed to strike further east and later in the day. No one had been prepared for its might!

So we listened and we tried to sympathize, and we tried to imagine what it must have been like on that terrible day in Cameron. "So I guess that explains why the town looks so brand-spanking new. Wasn't much left of it back then." John had finally got back around to where the discussion began. "Funny how right after Audrey, people kept sayin', 'I'm never comin' back,' but, sure enough, most of 'em did. What are you going to do when you got family here that goes back generations? When you got a livelihood, and that's the only thing you know how to do?"

The place became quiet as we all pondered that thought.

John was the next person to speak again. "Last call, guys...I gotta work tomorrow!"

I remembered then why we had come into the bar to begin with—besides to quench our thirst, I mean. "Say, John, we're looking for a place to stay tonight, but we're kind of on a budget. Anything you could recommend?"

"If you're looking not to spend a lot, why don't you ask that gentleman down at the end of the bar?" He pointed to a well-built guy, maybe in his forties, sitting by himself and nursing a whiskey. I hadn't noticed him until then; he must have just walked in.

"Who's he?" I inquired.

"The sheriff."

The looks on our faces must have told John that Paddy and I were a bit concerned at that last piece of information. He kind of smiled. "No, don't worry. Go ahead and ask him. He might be able to help you out."

Still a bit apprehensive, I got off the stool and approached the gentleman at the end of the bar. "How ya doin'? My buddy and I were looking for an inexpensive place in town to stay, and John, the bartender, said you might be able to help us out."

"I got an empty cell at the moment with clean cots...does that work for you?"

I tried my best not to sound too surprised, and in contrived matter-of-factness, I replied, "That certainly works for us if it works for you." I still didn't believe what I was hearing.

"Then, if you don't mind, let me finish my drink, and I'll get you boys set up for the night. Those your motorcycles out there?"

"They are..."

"They'll be safe!"

Ten minutes later, we were at the sheriff's office. "Here are your accommodations," he said as he unlocked the cell door. He asked us for our wallets and any other valuables. "I'm going to lock these up in the safe just as a precaution. Your door will be unlocked also."

We thanked him profusely.

"I'll be back here about six a.m....you boys drink coffee?

We said that would be real nice.

Before falling asleep, Paddy and I joked about whether we had unwittingly walked into some sort of weird legal trap by voluntarily putting ourselves in jail.

A few months earlier, the Supreme Court Miranda ruling had been rendered, which basically said that anyone arrested or charged with a crime must have his legal rights read to him, including the right to remain silent. Perhaps that, and the fact that we were now in a jail (plus that I had just finished reading some Franz Kafka and Joseph Heller's *Catch-22*), prompted my crazy dream in which Paddy and I were denied permission to leave the next day.

In the jumbled dream sequence, because we had voluntarily requested incarceration, we had waived our legal rights, and the friendly sheriff, still smiling but in a more sinister way, informed us apologetically that his hands were tied by the Napoleonic Code's guilty-until-proven-innocent thing that they had down there and that the state of Louisiana took vagrancy seriously and so, just walking out of jail—for the moment, at least—wasn't going to be that easy but don't worry our motorcycles would be safe while he tried to sort things out legally it might take a while.

But the next morning, the sheriff did arrive, not only with coffee as promised but also some beignets. We had never heard of such things, but they were better than any donuts we had ever eaten. We chatted a while, and I was feeling kind of guilty for having had that weird dream. We had just experienced southern hospitality at its finest, after all. So I never mentioned it, not even to Paddy.

After breakfast, our sheriff friend got our wallets out from the safe and walked us out to the motorcycles that were parked under a huge live oak decorated with Spanish moss, something we hadn't noticed in the darkness the night before.

Looking up at the tree, the sheriff said, "She survived Audrey. Maybe she'll live a couple hundred more years. You boys have a safe trip now, y'hear?"

Still Bird-Watching (on My Harley)

Once a bird watcher, always...they say. I've already discussed my ornithological proclivities, but I now make the claim that back in the early '60s, I was the only bird-watching Harley rider in the country (and, of course, that probably means in the entire world). OK, although I concocted that claim to fame rather unscientifically, I defy anyone to prove otherwise!

Again, *birder* still hadn't entered the lexicon, so I was a *bird watcher*. Not just a bird watcher who happened to own a Harley-Davidson, I actually got in some productive bird-watching while riding. That's right; while cruising at sixty, I managed to keep my eyes on the road *and* utilize peripheral vision to spot some pretty interesting avifauna. And riding sans helmet, which is what we did in those days, allowed for unobstructed visibility and greater opportunity to observe all sorts of wildlife. So, in my still-invincible adolescent mind, the advantages of no helmet far outweighed its sheer stupidity.

As I said, not just birds but other forms of wildlife would catch my attention while I rode. Like turtles. Each crushed turtle carcass on the pavement was upsetting to me. So much so that I would often pull off to the side of the road and, if necessary, wave down traffic to rescue turtles attempting to cross and carry them into the surrounding habitat. Back in the early '60s on a ride through upstate New York, I'd found a box turtle stranded in the center of a highway ducking his head in and out of his shell as cars and trucks sped by. "This is your lucky day," I'd said as I tucked him into my jacket and took him home. Peter (my mom named him) lived with us for several years before I finally relocated him back to the wild.

Seeing some unusual bird while riding often brought me to a complete stop. On one occasion along the Gulf Coast, I spotted an osprey flying above with a fish clutched in its talons and immediately pulled over for a better look. Ospreys had become quite uncommon in the Northeast due to the devastating effects of DDT use, but they could still be seen along the Gulf Coast. My excitement was met with indifference on Paddy's part, but I was used to that.

Another time, in the vicinity of the Texas-Louisiana border, I spotted a most extraordinary bird with an impossibly long, graceful tail perched on a fence post. This also brought me to an abrupt halt: my first scissor-tailed flycatcher! Especially spectacular is the male, with a tail actually greater in length than his body. And then consider its Latin name: *Tyrannus forficata*. The genus name *Tyrannus* shares its root with the more familiar *Tyrannosaurus rex* and means "tyrant-like." Like its other kingbird relatives, this fellow can be very aggressive indeed when defending his territory, often attacking larger interlopers like crows and hawks. Species *forficata* derives its name from the Latin word for "scissor," an apt description of its elegant tail. No surprise that it's also known as the Texas bird of paradise and was ordained the official state bird of Oklahoma. Seeing this bird served as another reminder that we were indeed entering exotic territory—or at least for me, it did. Certain things matter more to some than others—it's that simple!

While I dismounted my bike, the bird flew off to a tree maybe fifty feet away. I grabbed my binoculars, which were always kept within reach for just such occasions. A barbed-wire fence with a No Trespassing sign confronted me. Not just trespassing, but hunting, fishing, trapping, berry picking, and a host of other outdoor pursuits were disallowed, but none could be done without entering the property to begin with, so listing them all seemed pointless. In any case, for some reason, the word *trespass*—for me, anyway—had always carried the overtone of a stern biblical injunction of the "Thou shalt not!" variety, so I paused to see if any farmers, ranchers, or clergymen were standing watch. Seeing none, I whispered, "And forgive us our trespasses" before depressing the barbed wire and stepping onto the property. And in breaking the law, I managed to add a subversive quality to the nerdy pastime of bird-watching. But in fact, the illegal act did afford a much closer look at my quarry, revealing the magnificent, salmon-pink coloration under its wings, something I had missed before. Although these beauties would continue to appear with some regularity along the highways of Arkansas and Texas, they never failed to amaze me.

Later in the trip (we're still on the subject of tough birds), somewhere on a desert highway north of Monterrey, Mexico, a roadrunner dashed across the road (fittingly enough) right in front of us. How tough is this bird? This feathered sprinter regularly takes on rattlesnakes and devours them for lunch, that's how tough! "Did you see that?" I yelled to Paddy for confirmation.

"Yeah...what was it—a roadrunner?"

"You know that bird?" (I was, frankly, quite astonished.)

"I watch a lot of cartoons!"

The fact that he could actually identify the bird was a tribute to the role of Looney Tunes in furthering our appreciation of the natural world. "I wonder if Wile E. Coyote ever catches the Road Runner," I mused.

"On TV, or in real life?"

"In the cartoons!"

"Don't think so," said Paddy. "If he did, he would kill the Road Runner, and that would be the end of that."

"There's always Woody Woodpecker."

"Woody Woodpecker?"

"In the category of bird-related cartoons I'm talking about," I said sagely.

"Well, in that case, there's also Heckle and Jeckle, the crows…"

"They're magpies."

"Magpies?"

"Magpies. Not crows!"

"You sure?"

"Pretty sure."

Paddy insisted, "None of them are as good as the Road Runner. So let's hope the coyote never gets him!"

And so ended another inane conversation.

Vultures too would frequently be seen soaring their lazy circles above us, and we were well aware of their reputation as harbingers of ill fortune. "You think they know something?"

"Always waiting…always waiting."

That evening, we splurged for a five-dollar-a-night motel. Soon enough, I was splayed out on a raggedy-ass, sagging mattress, thinking, *This is going to be a long night.* So I reached into the nightstand, only to find Gideon's Bible, which I perused by the meager illumination of a thrift-store table lamp. A few selected chapters from scripture would be the perfect elixir for that pent-up feeling that came from a long day on the road. Especially soporific were verses in Genesis that droned on and on about who begat whom, but on this particular night, Leviticus, chapter 11, verse 13 caught my attention. "Paddy, you remember that large bird we saw this afternoon with the fish in its talons, the osprey?"

"Vaguely."

"Vaguely? How do you possibly not remember a large bird of prey with a fish in its talons?"

"I can see you're going to tell me about it whether I'm interested or not. So, what about it?"

"I can tell you tomorrow."

"No, go ahead. Now that I'm awake…"

"I didn't know you were sleeping."

"Go ahead. What about the osprey?"

"It says here in the Bible that it's not kosher…to eat an osprey is an abomination."

"It's not kosher? You woke me up to tell me that an osprey is not kosher?"

"I said I thought you were still awake."

"You're going to read the Bible to me at"—he paused to look at the Big Ben alarm clock on the dresser—"eleven o'clock at night?"

"Just the part about the osprey. Here's what it says: 'These are the birds you are to regard as unclean and may not eat.' Then comes a whole bunch of birds, including the osprey!"

"Which explains why you can't get osprey on rye in a Jewish deli."

"For years, I've been wondering about that…"

"Will you please go to f'n sleep!"

The Fouke Monster

Starting in the 1890s and continuing through the early twentieth century, thousands of towns in the United States took steps to forbid black folks from living inside their borders. They were known as "sundown towns," so named because of signs posted at the city limits: "Nigger, don't let the sun go down on you in [name of town]."

Fouke, Arkansas was then such a town, something we learned during our time in Texarkana, its bigger neighbor down the highway. We actually learned two other things about Fouke: that it was the self-proclaimed toughest town in all of Arkansas and that it was being terrorized by a "bigfoot"-type creature that roamed the surrounding pine woods and bottomlands.

As far as the sundown-town phenomenon, the reader should bear in mind that much of the United States followed exclusionary practices then, not just the South. Towns and villages, country clubs, and other entities had written and unwritten policies barring Jews, "Orientals," "Colored," and pick your own persona non grata of the moment. College Point, Queens, where I hailed from, had no blacks, and I was told that this policy had actually been incorporated into the village bylaws, although I have yet to uncover documentation to that effect. The South, however, made no secret of its racial policies. Perhaps the openness of the bigotry was more shocking than the bigotry itself. That, and the fact that they meant business down there, as evidenced by the lynchings still taking place. Strange fruit indeed hung from those big southern trees.

Why we wound up there to begin with had to do with the fact that Paddy had served in the army from 1964 to 1965, and much of that time he had spent at the Red River Army Depot in Texarkana. While on our wanderings, why not pay a visit to one of his old army buddies—Roy was his name—and why not renew acquaintance with an army wife he knew whose husband was overseas?

Paddy still had connections with the quartermaster on base, through whom I secured most unique lodging: the building in which the mattresses were stored. While my friend pursued his liaisons, I was sleeping comfortably on top of a stack of mattresses ten deep.

During our stay at Red River, I was introduced to Paddy's army buddy Roy, who came from the town of Fouke just down the road. We soon discovered that its reputation for being the toughest town in Arkansas wasn't entirely braggadocio. On a ride to a Harley dealer in Shreveport, Louisiana, we stopped to refuel in Fouke, our very first time in that town. An unusually large crowd of men was milling about at the station, and when we pulled up to the pumps, we saw why: a brutal fight was in progress. The combatants were local boys in their twenties and would have been equally matched were it not for the fact that one of them was wielding a wooden hatchet handle. The opponent, though still on his feet, was covered in blood. As unfair as the contest was, no one intervened. Paddy and I remained with the bikes and observed from a distance, as neutral as Switzerland.

In a matter of minutes, the bloodied man lay prone, and although he was still conscious, the fight was over. The victor and the rest of the crowd slowly dispersed in pairs and small groups murmuring among themselves. I overheard a fellow say to no one in particular, "Well, that ol' boy certainly learned a lesson today!"

Still at the pumps, Paddy and I finally got around to filling our tanks. "What was that all about?" I asked the young gas-station attendant.

"A girl."

We asked no further questions, kick started our bikes, and continued on to Shreveport.

Now, about the local Sasquatch: the previous day, I had come across an outdoors-supply shop, one of those little independently owned places specializing in hunting, fishing, and camping supplies, mostly army surplus, that would soon be forced into extinction by the advent of Cabela's, Bass Pro Shops, and the like. I needed a new flashlight and while there struck up a conversation with the owner about local deer-hunting regulations, curious to see how they compared to those in New York. Here's where I first heard about the Fouke Monster.

The store owner said that he himself had never actually seen the creature, although he had been hunting here all his life. But he knew of some folks that had—or maybe heard it at night. "And some of those folks are quite reputable," he added. "One ol' boy got a pretty good look at it from his tree stand right around dusk. There's also a family living way off the highway says the thing broke into the chicken coop at night, and they let loose with both barrels of the shotgun to chase him away. Got a lot of folks around here worried!"

"What's it look like?"

"There's general agreement that it's big, about eight foot tall. And it's got a deep, loud cry of sorts…some say it's got a bad smell."

I paid for the flashlight and said it had been nice talking.

Erwin "Erv" Krause

The subject came up again the following night. I had been riding around, sightseeing, and when I returned to the mattress room later that afternoon, there was a note from Paddy:

Erv,
    We got dates tonight with Roy's sister and her girlfriend. Roy says it's OK to use his Chevy. I'll be back here at 5:30.

Guess who?

P.S. Roy says they're real ugly! Kidding!

Paddy showed up at six with Roy's '53 Chevy Bel Air ("Sorry I'm late…it's a long story"—which I never did hear, but it was obviously something to do with a lonely army wife), the one with a three-speed column shift and spongy bench seats front and rear…great ergonomics for submarine-race watching or a drive-in movie, which is where we were going that night to see *Spartacus*, with Kirk Douglas (whom I was told I resembled, although I couldn't see that despite having the same flattop haircut with the sides long. I loved that movie and in fact had already seen it several times. But did guys really have flattops in 71 BC?).

    Paddy had given Roy his bike in exchange for the Chevy that night. At the end of the evening, we would return Roy's car to him. A fair-enough arrangement, I thought.

    Roy's family lived in a farmhouse out in Fouke, and his youngest sister, Sharon, still lived at home with the parents. The house was at the end of a dirt road about a mile off the paved highway. Mom and Dad met us at the door and invited us in. The place looked real country, and so did the folks. There was a nice ten-point buck mounted on the wall, a gun cabinet in the living room, and crucifixes everywhere. I hadn't gotten my hopes up that we would actually get lucky that night, and the Christian décor didn't change that outlook, although Roy's sister Sharon and her friend Rebecca ("Call me Becky") certainly were cute in that country sort of way.

Mom and Dad had us sit down at the kitchen table covered by a calico tablecloth, where we drank sweetened ice tea while the girls disappeared to fine tune their makeup. To strike up a conversation, I complemented Dad on that nice set of antlers in the living room.

"Got that deer four or five years ago," he said. "We haven't had much luck for the past few seasons...some of the boys are saying it might be on account o' that monster...you heard about him?"

I told Roy's dad that I had just found out about the monster the previous day from the guy at the sporting-goods shop. "Have you seen him?" I asked.

"Not personally, but my cousin Earl knows a fellow who actually did spot him in some bottomland not far from here. Fellow was draggin' a buck out of the woods, and the thing was following him. That ol' boy got so scared, he let the deer lie and run out of the woods to where his partners were, and when they went back in with flashlights, the deer was gone!"

We didn't have time to discuss it further; the girls were ready to go. We said our good-byes, walked to the car, and, like real gentlemen, opened the doors for the girls. Paddy got behind the wheel, and I got in the back seat with Becky. A final wave to the parents on the front porch, and some fatherly words from Dad: "Y'all be safe now, y'hear!" *Funny,* I thought, *he didn't say, "Drive safe."*

So, off we were to the drive-in to see *Spartacus,* a great date movie with plenty of action for the guys and romance for the ladies. Once we arrived, Paddy parked toward the back end of the parking area even though plenty of spaces were available up front. Good move, I thought. We bought the usual American junk food and soda at the food concession and settled in for the show.

We actually did get to see most of the movie, but those country girls were not as prudish as I would have guessed. We stopped making out for the gladiator scene where Kirk Douglas battles that ripped black dude Woody Strode, in which Strode spares the life of Douglas despite all the thumbs-down gestures from the Romans, and then Strode himself is speared by a Roman soldier. After that scene, Paddy and I got into a brief debate about gladiatorial skills and who would you rather be, the guy with the sword and small shield or the guy with the pitchfork and the net? The girls decided it would be a good time to powder their noses.

When they returned, it was more of the same: intense making out and intermittent movie watching. The making out was again put on hold for the final scene in which Douglas, the leader of the slave rebellion, is crucified along with his followers. The girls needed some time to dry their tears before we headed back to the farmhouse.

On the way home, we made a little detour to a de facto parking area off the dirt road that led back to the house. The fact that the stop had come at the girls' suggestion got my hopes up. The place was surrounded by woods, and, once the motor was shut down and the headlights off, insanely dark. So dark that even making out with eyes wide open, you weren't going to see much anyway. This would be to our advantage, since the sexual revolution hadn't really gotten started (especially in Arkansas), and girls were still somewhat modest back then. Things were going hot and heavy when it happened—not the sexual revolution, but a shrill scream from Sharon up in the front seat with Paddy. "He's there...he's there! I saw him! His ugly face was pressed against the windshield!" This was followed by the plea, "Start the car! Let's get out of here. Take us home, take us home!"

By now, Becky and I had propped our disheveled selves up in the back seat, she doing some panicky wardrobe adjustments and me wiping the fogged-up windows to see if anything really was out there. "I don't see anything," I said. "Lemme go out and take a look!"

"Please don't! Keep the doors locked. Start the engine, and let's just get out of here!" the girls implored.

Paddy cranked the starter and turned the headlights on. "Let's see if we can catch his eyes in the lights," he suggested.

By now, the girls had worked themselves up to a frenzy and in unison pleaded, "Just go home. Drive us home!"

Which we did.

Postscript

Seven years later, in 1973, I was reading the entertainment section of *The Village Voice* and came upon a review of a B movie entitled *The Legend of Boggy Creek* that was currently playing at a small downtown art-movie theater. The review held little interest for me until I got to the part about Fouke, Arkansas! The movie was about the monster, the creature Sharon had seen through the steamed-up windows of Roy's '53 Bel Air on that hot summer night so long ago. They had actually made a movie about the brute!

An enterprising salesman from Texarkana, Charles B. Pierce, had borrowed $100,000 from a local trucking company and hired locals, including many high-school students, to make the ninety-minute film using an old 35-mm movie camera. That film would eventually become a cult favorite and earn $20 million in box-office revenue—not a bad return on Pierce's investment!

Did I go downtown to see the movie? What do you think?

Bury My Heart in Texarkana

Texarkana...is it Texas or Arkansas? The name should give you a clue: it's both. One of those quirky geopolitical anomalies in which the main drag, State Line Road, actually forms part of the border between the two states. Such boundaries, unless there's a river, body of water, mountain range, or other geographical feature, can be rather arbitrary, and on less traveled roads, you can cross over into another state without even realizing it, because no one takes the time to put up a welcome billboard. Back in 1966, however, a distinct difference could be seen between the opposite sides of State Line Road: on the Arkansas (east) side, it seemed that every other business was a bar or saloon, but there was not one such establishment across the street in Texas. That's because after our national Prohibition was repealed in 1933, Arkansas decided to once again permit alcoholic beverages, while Texas voted to stay dry.

It was on a hot July afternoon that I found myself riding down State Line Road with a thirst that only a cold beer could quench. All sorts of creative beer signs, like neon sirens, lured thirsty Texans and at least one thirsty New Yorker to visit the state of Arkansas.

Like many other towns west of the Mississippi, the broad main drag of Texarkana called for angle parking, so I dutifully positioned my Harley into an available space (taking the precaution to face the machine out in the direction of the street) and soon found myself in the climate-controlled confines of a local drinking emporium. Looking over to the teetotaling side of the street, I wondered, who the hell were those puritanical, Bible-thumping moralists who had decided that an occasional beer was somehow contrary to God's word? What verse in the Bible actually proscribes the consumption of alcohol (a by-product of fermentation, a process brought about by yeast—which God himself created [if you accept Genesis])? And didn't Jesus himself turn water into wine (and not the other way around) at the Feast of Canaan? These thoughts imparted a sense of political protest to the simple act of quenching my thirst…in my mind, anyhow.

From previous experience, I knew that a cold beer paired well with Patsy Cline or maybe Ray Charles—or any country music with that high, lonesome sound. So I invested a quarter in the Wurlitzer before getting comfortable on the barstool. Next to me sat a lean, darkly complected gentleman wearing a snap-button Western shirt: white with an indigo Navaho geometric pattern. *I should get me one of those*, I thought. "Where y'all from?" Perhaps he had detected an eastern accent of some sort when I'd requested jukebox quarters from the bartender.

"New York."

"Not Vermont?" (This he pronounced more like "*Ver*-mont.")

So he had seen me pull up on my motorcycle. But I really had no desire to get into the story behind those Vermont plates. "Oh, those plates on my bike…that's a long story." But I did go on to share with him that a friend and I had been on the road for a while, and we were stopping over in Texarkana so he could meet up with some old army buddies (I left out the tale of Paddy's little liaison) before making our way to Mexico. The conversation continued in this vein about places we had seen and our experiences along the way.

Escape from the USA

Somewhere during the conversation, he looked closely at me for a few seconds and asked, "What are you...I mean, what nationality?"

"German...both my parents were born there, but I was born in New York."

"I thought so," he said, rather self-congratulatory. "My sister married a German guy, and he kinda looks like you."

"What about you?"

"Me? I'm one hunnert percent American—*real* American. Mom's Choctaw, and Daddy's Cherokee."

The conversation then took a turn to all things Indian: about the differences between the tribes, about life on the reservation where his grandparents still lived, and so on. I listened while he did most of the talking. I learned that his name was Will Hawksbill—most people called him Hawk—and that he lived in Oklahoma with his wife and five-year-old daughter. I guessed he was in his midthirties. I learned that he drove a truck for a large shipping firm and was in Texarkana overnight to pick up a new rig for his boss. I shared with him that I had recently graduated college, had been accepted into a graduate program at the University of Illinois, and was planning to go into education. The motorcycle trip, I told him, was kind of a last fling before joining the real world. Funny, isn't it, this propensity we have to share our private lives with complete strangers...people we'll probably never see again. And maybe that's exactly the reason we're willing to be so open.

This conversation, though, always seemed to come back to the subject of Indians. At some point, Hawk looked at me rather seriously. "Erv, what do you know about the Trail of Tears?"

I hesitated. In fact, I did know a few things about this tragic story, and not because of any history classes in either high school or college. It was actually a friend of mine with an interest in

Native American culture who had told me about the Trail of Tears, and I subsequently had done some more reading on my own. This was quite recent, and my knowledge was cursory at best. And quite honestly, I hesitated for another reason: would I be taking a risk by getting into this topic? I was experiencing some of the same uncertainty that I would have felt if a Jew (and a relative stranger at that) had asked me, a German American, what I knew about the Holocaust. Was it collective guilt, so to speak, making me ill at ease?

I did a quick risk assessment and decided that it was safe to answer Hawk as directly as possible. We had met just a few beers ago, but I sensed his intelligence and character, and I concluded that his question represented nothing more than an effort to continue a dialogue about something clearly important to him. After all, it was Hawk who had brought up the subject.

"The Trail of Tears," I began somewhat uncertainly. "Now, correct me if I'm wrong, but didn't that take place back in the early 1800s when Andrew Jackson, who was president at the time, evicted the Cherokee people from their ancestral homelands in Georgia and North Carolina and ordered a forced march to resettle them in so-called Indian Territory in Oklahoma? Wasn't most of their original land sold to white folks and converted to cotton farms?"

I went on a bit more with the little information I had, and Hawk listened patiently, supplementing my story with additional points. Lots of them, in fact. "Yes, it was Andrew Jackson, that no-good bastard—pardon my French—who was behind legislation called the Indian Removal Act of 1830. And it wasn't just the Cherokee. They were actually the last tribe to be removed in 1838. Other tribes were also forced out: the so-called Civilized Tribes, like the Choctaw, the Creeks, and the Seminoles. They were referred to as the Civilized Tribes because for the most part, they had adopted the white man's ways, becoming farmers and merchants, forming their own schools, wearing white-man's clothing…they even became Christian! They did all the right things, and still it wasn't enough. All my people were forced to travel during a terribly cold winter…thousands died."

Hawk went on at length, detailing this atrocity of ethnic cleansing (of course, he didn't actually use that term, which only entered our vocabulary during the 1990s Balkan crisis) perpetrated by Americans of European ancestry against its indigenous peoples…a difficult story to hear.

As I listened, not only did I feel shame that "my" people had been responsible, but anger at having learned so little of this (had I learned anything at all?) in my formal education…anger over an education that in some important way had failed me, and anger at teachers who had seen their role as passing along the "glorious history" of our country and shielding students from the unsavory bits.

Hawk had one more important thing to add. "And you know what really galls me? White folks have this expression, 'Indian giver,' for folks who give something away and then take it back. Indian giver! This, coming from people who broke just about every treaty they made with the Indian…over and over again making promises and then breaking them!"

An awful silence ensued, which thankfully was broken by Hawk.

"Jeez, I don't know about you, Erv, but I am awfully hungry. You like steak? Let me get us a steak dinner!"

"No, man, you don't have to do that…"

"You're right; I don't *have* to do it. I want to! Do you know that you're the only white guy I've ever met who knew anything about the Trail of Tears? You and me are having steak, and it's on me!

"But…"

"And I don't care how much you knew or didn't know on the subject. Doesn't matter. Now, let's get us a table."

And so we did.

Over dinner, the conversation continued, but mostly about other things and not so much Indians. When we finished, Hawk and I returned to the bar. "One for the road?" The bartender served up two Lone Stars. Hawk gave him a fifty. "Take everything, including the dinners, out of this."

We were rung up at the register, and the bartender returned a twenty, a ten, and a bunch of singles.

Hawk picked up the twenty-dollar bill and stared at it for a few moments before uttering, "Andrew Jackson, you bastard!" Then he called the barkeep back. "Could you break this twenty for me?"

"What d'ya want…singles, fives, tens?"

"Doesn't matter, doesn't matter. Just break the damn twenty!"

Arriving in Corpus Christi

East Texas was turning out to be a huge disappointment. The little we saw of Houston left us with the impression that it was a shithole: ugly, squat architecture, squalid industrial and commercial areas, and hot—hotter than we'd ever imagined. I knew it was Texas in July, but couldn't it at least cool down a bit in the evening? Apparently not. So we decided to take Route 35, hug the Gulf Coast, and make our way south to Corpus Christi. Maybe offshore breezes would help. And certainly, a city named after Jesus would offer relief of some kind.

We had heard about Padre Island, that long strip of sand east of Corpus Christi extending south to the Mexican border. Locals told us it was real pretty, with the best beaches in the entire state. A few days on the Gulf of Mexico might be just what the doctor ordered.

The journey had taught us a few things—one being that those musty, old hotels and motels in downtown areas and away from the new interstates offered a better bang for the buck. Although we sought lodging only once every few days, it still had to be economical. A plethora of bargain-priced lodging could be found on "Main Street, USA"—places I would later describe as "in the heart of town and on the outskirts of society." True, the rooms often harbored a stale-beer-and-tobacco fragrance that traveled with you after checkout, but such minor unpleasantness surely made for affordability. So, in a forlorn part of Corpus Christi, we came upon a cluster of compact, concrete-block structures called Beach Rose Cottages, the absence of any nearby beach notwithstanding. They seemed to meet our stringent criteria.

We knocked on the door of the only freshly painted unit, the one with an old anatomy-text style hand with finger pointing to the word *Office*, secretly hoping that the remodeling effort didn't mean a rate increase. The manager (or was it the owner?) came out in a cloud of tobacco smoke to greet us. Slightly built and with a deeply creased face, Cal was one of those guys who probably looked a lot older than he actually was. I couldn't actually state that as a fact. He was a real chatterbox. We immediately learned much of his life history: he had formerly worked in the shipping industry, had been divorced twenty years earlier, and hadn't touched a drop of liquor in that same period, clearly implying cause and effect but taking it no further. He got down to business, "Got a couple units left."

Paddy and I surveyed the premises and got the distinct impression that more than just a couple units were left, but conceded that it being midday, maybe guests were at the beach. Our hopes remained high for some "off-season" rates, since it was Texas in July.

"How long you boys plannin' on stayin'?"

We kind of looked at each other and told him we weren't sure—it depended on how we liked it.

"Sure is nice, beholdin' to no schedule…lemme show you what's available. By the way, do you boys smoke?" he asked, pulling a Lucky Strike package from his shirt pocket.

"No…and no."

"Wish I'd never started," he confessed in that way older folks had of advising you not to go down a certain road. "Too late now. Hard to break habits built over the course of a lifetime."

Cal showed us a unit that was actually a lot nicer on the inside than its in-need-of-a-fresh-coat-of-paint exterior suggested. No beer or tobacco smell, but that musty odor masked by Lysol we had gotten used to by now. On the plus side, there was a small kitchen area with a stove and fridge. "How much?"

"How 'bout fifteen dollars a—"

"Whoa...we're not exactly Rockefeller."

"—a week?"

"Of course—a week." I tried to disguise my incredulity. "Whaddaya think, Paddy...fifteen work for you?"

There was a contemplative pause. "Works for me, I guess," answered my friend, trying his best to sound like we were taking the rate under protest (and maybe overdoing it just a little).

At fifteen dollars a week, there was certainly no compelling reason to haggle. We didn't even have to stay the whole week to get our money's worth—which turned out to be a good thing, because it didn't take us long to wear out our welcome. It actually happened the following day.

"Corpus Christi" Means "Body of Christ"

We awakened with a plan to visit the beautiful barrier island of Padre Island we had heard so much about. We packed some sandwiches and Coke cans that by mistake had been left in the freezer compartment overnight. We expected they'd be thawed out but still nice and cold in a few hours. Our thin motel towels would have to do for the beach.

Gulf of Mexico waters are unlike anything on Long Island. That dark blue Atlantic off Jones Beach seemed cold even in August, and there were two accepted methods for entering: the slow way—inch by inch until the waters were up to your chest, followed by a final, complete submersion—or the more heroic, get-it-over-with-quickly method, a sprint ending in a headfirst plunge accompanied by a gasp and a Tarzan-like yell. Getting into the turquoise Gulf waters off Texas was akin to slipping into a lukewarm bath, the difference between body temperature and sea being minimal. Beach sand here was also softer, with a talcum-like feel underfoot, making the whole experience exotic.

Even more exotic was the presence of two lovely girls laughing while setting up their blanket nearby. The beach was surprisingly unpopulated, yet another difference from Jones Beach. It wasn't long before I realized that between laughter and giggles, the girls were speaking German. I summoned up the courage to approach them with a few words in their language.

Escape from the USA

"Guten Tag! Ich heisse Erwin, und ich kann ein bisschen Deutsch sprechen. Wo sind Sie von?" (Good day. I'm Erv, and I can speak a little German. Where are you from?) The girls immediately became very friendly, either because they were astonished to hear their language being spoken on a remote beach in Texas, or simply because, as we all know, German is the language of love.

Our new friends Heidi and Renate, from Hamburg and Bremen, were employed on a German freight vessel, currently docked. They had a few days to themselves before embarking again. Their English was about as good as my German, which is to say quite rudimentary, but by transitioning frequently between languages, conversation flowed remarkably well, and my friend could participate.

The day flew by. At one point, we played "knights on ponies" in about four feet of water. With girls mounted on your shoulders, the object is to knock your opponents into the water—an aquatic jousting match. Each round of jousting ended in hysterical laughter. With a girl's legs wrapped around my neck, I always found this game one of the more erotically charged beach pastimes, right up there with applying suntan oil on the soft, exposed backs of the fair sex.

We talked about our travels, and the girls showed considerable interest. They had never been to New York and had many questions about the City. They also shared much about their voyages and favorite places. I guess you could say we hit it off quite well.

It didn't hurt that the girls were both attractive and athletic. Their long hair was at opposite ends of color spectrum: Heidi was a flaxen blond, classically Northern European, and Renate was an impossibly dark brunette, almost raven. They didn't even realize how beautiful they were, which made them all the more so. Side by side, the contrast was stunning. At that time, my flat-top-style hair was blond and sun bleached, which might explain my gravitation toward Renate, though it couldn't be said that we actually had paired up in any way.

As our beach day came to an end, we asked the girls how they planned to return to their ship. They had arrived by taxi, they said, and would use the phone booth at the concession stand to call for a ride back. "Forget the cab," we said. "We'll take you back!" They happily agreed.

Back at the dock, the girls invited us for dinner that evening. They both worked in the ship's galley, and a nice dinner would be no problem. Needless to say, we accepted the invite and then departed for our rental cottage to shower and freshen up, filled with great expectations for the rendezvous with our Rhinemaidens.

Our motel was about twenty minutes away, and the return trip took us through some sketchy neighborhoods—not surprising for a port city. Although uncertain of the route, we figured that as long as we stayed near the bay, we couldn't get too lost, since our little cottage colony bordered the waterfront.

Just a few miles from our destination, we were proceeding slowly on a street going through a marginal kind of commercial area with apartments on the second level and run-down old houses scattered about. Slowly turning at an intersection, we passed a group of five kids in their late teens or early twenties hanging out in a small parking lot next to a run-down wholesale plumbing-supply business. Seeing us, they yelled, "Whoa…hey! Stop a minute. We wanna talk with ya!"

It was not uncommon for locals seeing two motorcyclists with out-of-state plates to want to strike up a conversation. The usual questions were, "Where ya from? Where ya goin? How long ya been on the road? Any problems with the bikes?" And so forth.

Escape from the USA

Besides being curious, many were envious of our adventure. I recall one fellow working in a gas station back in Mississippi saying, "Man, you guys are lucky...someday, I'm going to do this!"

And so, I said to Paddy, "Pull over. These guys want to talk with us."

So we did, and while we sat there with engines idling, not bothering to even turn around, out of the corners of our eyes, we could see them jogging toward us. Then came the unexpected. Without warning, fists came raining down on our heads (remember—no helmets!) and backs. Stinging, hard punches. The first one hit my head so hard, I was certain the deliverer of that blow had broken his hand. No time for kickstands. We dropped the bikes, engines running, and rolled over and up to our feet, retaliating as best as we could, until in ten seconds or so, there was kind of a standoff.

Paddy and I were now facing this menacing semicircle, focusing on the guy who appeared to be the leader, as evidenced by the six-inch blade now pointed at us. At the same time, we had to be prepared for an attack from behind, so with clenched fists, we assumed a defensive crouch, constantly pivoting our feet to ward off a renewed assault.

"What the hell is this all about?" I demanded. "What do you guys want?"

The fellow brandishing the knife was clearly in charge. A rangy, "Anglo"-looking guy (we had recently learned that in these parts, the term *Anglo* was the designation for Anglo-Saxon types as opposed to those of Mexican ancestry) made it clear what this was all about. "You know what...you think you can come down here and mess with our girlfriends?"

"What the hell are you talking about? We were at the beach all day. Got nothing to do with your girlfriends!"

"That's not what they said. They said two guys on motorcycles from out of town were messing with them, and they told them to go to hell, but these guys kept sayin' shit!"

I continued to protest. "Look, I don't know what went on with your girlfriends, but I know it wasn't us, because we were at the beach all day. There must have been somebody else messin' with them!"

During the heated exchange, one of them, a big, fat Mexican guy to my right, like some diabolical cheerleader, was doing his best to heat things up. "Don't believe them, Tommy. They're lyin'. Let's just cut'm up. Cut'm up!"

I was still trying my best to convince them that we had had nothing to do with their girlfriends. "Look, it couldn't have been us. We were on Padre Island all day. We were nowhere near here…"

Throughout the verbal exchange, the fat Mexican continued to plead with the knife-wielding Anglo, "What are we waitin' for? They're lyin'. Just cut'm up. Cut'm up!"

There's a phenomenon that every college freshman learns about in his or her introductory psych course known as the flight-or-fight response. First described by Walter Bradford Cannon back in 1929, the theory states that all animals react to perceived danger with a sympathetic-nervous-system response that prepares them for either fighting or fleeing. To facilitate violent muscular action either way, hormones induce an increase in heart and breathing rates, blood flow to the muscles, and muscle tension.

Where it gets a little complicated is exactly what is behind the actual cognitive decision to flee or to attack? The body has been prepared for either one, so which do you choose, and why? The choice partly relates to the individual's perceived control over the situation. However, perceived control may not reflect reality. The same can be said regarding one's cherished beliefs about his or her actual fighting prowess.

Back in the streets of Corpus Christi that hot July afternoon, I was not reflecting on what I had learned in Psych 101—at least not consciously. What I do recall, however, is the revulsion of envisioning myself bleeding to death in the gutter with a fat Mexican standing over me yelling, "Cut'm up…cut'm up!" and then dying in a final burst of rage at the indecisiveness that led me to my fate.

So I made a decision midsentence. I think I was saying something like, "We were nowhere near—" and my left fist shot out and landed flush on the Anglo kid's jaw with a force that could be felt up the length of my arm.

He went down almost in place. They call it the "sweet spot"— like the location on a tennis racket or baseball bat where contact with the ball propels it with the greatest velocity. An athlete feels it instantly. In boxing, also, there's that punch. Most of the great boxers had it, like Robinson and Marciano. Even some who never approached greatness, like Ingemar Johannson, wielded "the Hammer of Thor" that was his right hand. For southpaws like Carmen Basilio, it was the left. Not those big, loopy, cinematic, John Wayne, cowboy-barroom-brawl punches. The real punches were such that if you weren't watching closely (and even if you were), you would say, "What just happened?" Often, it took several replays on TV to reveal that punch's mystery. Don't misunderstand. Not in my wildest fantasies could I come close to any of the aforementioned pugilists. But, on the other hand, if Mack the Knife had still been upright after my best punch, I would have been in real trouble!

But he did hit the ground. And I immediately directed my fury toward the fat cheerleader, who was already backing up. As a result, my first punch at him didn't do any damage, but

I continued swinging. Out the corner of my eye, I could see Paddy covering up and in sporadic bursts winging away at the other three, who were all over him. You'll recall that he was a strong boy who could dead-lift VW Beetles, so I was hoping he could handle those punks until I was finished with my endomorphic tormentor.

But things weren't going so well on my end. Although my rage was such that if he had died at my hands, I would have experienced no remorse, the outpouring of emotion now became a hindrance, and I was running out of steam. I was crazy, though. "Cut him up? Cut him up?" I screamed his own words back at him. In a few seconds, I had him trapped in a debris-filled corner of an empty yard. He went down to his knees, not so much because of my punches, which were losing their effectiveness, but from ducking under them.

Then came a surprise when he sprang back up to his feet in a batter's stance with an old two-by-four he had picked out of the rubbish. I backed off to put some distance between us and quickly weighed my options. Maybe a quick feint. He would then swing and miss, and I could tackle him before the next swing. In hindsight, this is what I should have done, but in my eagerness to come to my friend's aid, I recklessly bull-rushed. He landed a chopping blow to the back of my neck that sent shock waves down my spine. Momentum, however, allowed me to complete the tackle, and, oblivious to the damage he had inflicted, renewed the offensive with a barrage of arm-weary punches, most of which were absorbed by his own fat arms.

At that moment, a man in a white tank-top undershirt leaned out from his second-story window and yelled, "Get the hell out of here! The police are on their way!" Almost instantly, our attackers ran off, including Mack the Knife, who had come to and staggered off with his associates.

Escape from the USA

I ran over to Paddy. We put the bikes back on kickstands and then beat the dirt and gravel off our bodies. "You all right?" I asked, seeing that he was covering his eye with his hand.

"I don't know...how do I look?" He slowly removed his hand to reveal an angry welt under his eye with the potential to get even angrier. He then began testing his jaw by slowly opening and closing his mouth with the help of his hand. I saw the beginning of a smile, which told me nothing was broken. "My neck feels twice its size," he said. "What does it look like?"

I turned my back to him to get a better look. "A little reddish, but no blood."

The bikes started up OK and had suffered no apparent damage. With engines idling, we looked up toward the window where, moments before, a good Samaritan had yelled the warning that had effectively halted the brawl. But it was shut, with curtains drawn. No sirens were heard, and no police were showing up. That guy yelling out the window had done us a huge favor.

"Let's just get the hell out of here!"

Back at our room, we took lengthy showers to cleanse ourselves of the mix of Coppertone, sweat, and road dust from the brawl. The warm water was therapeutic but stinging, revealing abrasions we hadn't even been aware of. My neck was so stiff that turning my head became impossible. We sat at the edge of our beds with ice from the fridge applied to the body parts most in need.

But we did have a date that night.

So we splashed on some English Leather, slipped into clean dungarees and collared shirts that had been stowed in the bottoms of our saddlebags for occasions such as this, infrequent as they were. Surely I'm guilty of anthropomorphizing, but the bikes both started on the first kick, perhaps sensing that we had already had our fill of difficulties for the day, and we set off for the docks where our fräuleins were waiting.

The evening was *ausgezeichnet*. The girls prepared a delicious German-style dinner with plenty of cold bottles of Beck's in the fridge. We needed to explain the flamboyant welt under Paddy's eye and the wincing initiated by each slight turn of my head, so we told them about the donnybrook while they listened incredulously and sympathetically. Talk turned to the amount of violence that seemed so characteristic of America, but we assured them that such random attacks were quite uncommon—although this was the South, after all. The conversation then took a more pleasant turn and flowed smoothly for the rest of the evening, once again alternating between German and English. We shared more about our families, and, just like at the beach, there was much laughter. Before our departure, we made arrangements for the following night, maybe a movie.

Both of us were feeling good on the ride home. The time spent with Heidi and Renate had made up for the insanity of the afternoon, and our spirits were buoyed by prospects of yet another night together. At the entrance to the cottages, the flickering neon sign swarming with insects announced, "…each…Ro…Cott…" Funny that we hadn't noticed it the night before. The cottages were dark except for the flickering glow of a TV coming from Cal's place. No sooner were the kickstands down when he came out with his omnipresent cigarette glowing. "You fellows have a little run-in with the McCaskill gang earlier today?"

Oh, shit, I thought. *Now what?* "I don't know who they were, but yeah, we did get into a fight with some guys who thought we were someone else," I replied. I thought the whole incident was behind us.

Escape from the USA

"Well, about an hour ago, three cars pulled in here loaded with them fellows carrying bats and chains. They told me to tell you they know that you're stayin' here, and they're comin' back to finish the job."

My thoughts suddenly filled with an apocalyptic vision of three cars disgorging a cargo of toothless, tobacco-chewing rednecks leering at us with hundred-proof, gap-toothed smiles and gripping bats and chains with heavily veined, sinewy hands, all bearing an eerie, inbred resemblance and united by that code of justice in which the honor of the clan is supreme. Mack the Knife was there, along with his amigo El Gordo pointing a fat finger at us and yelling, "What are we waitin' for!"

With their collective intelligence on the far left of the bell curve, reasoning with them was not an option. Their lust for revenge would not go unsatisfied. Later in life, I would safely encounter such characters on the pages of Cormac McCarthy novels or watching a Sam Peckinpah movie, but at this moment, the vision was far too real!

Paddy and I looked at each other and knew we needed to get the hell out of there, and in a hurry. I still couldn't believe what was happening. This crazy day that had started out with such promise and then took a downward turn, only to bounce back so beautifully, had now turned back into an even worse nightmare. The flight-or-fight response had returned with a vengeance, only this time, there would be no fight! I think I heard Paddy mutter something like, "Feets, don't fail me now!" before we got to stuffing belongings back into our duffel bags, shaking so much that packing and bungeeing became a struggle. But pack we did, and in no time, we were set to go, hoping nothing had been left behind.

One last thing was a quick chat with Cal. I told him we were heading back north, thanks for everything. The part about heading north was in case the gang came back and asked in what direction we had gone: the old misdirection play. They would go north—a clever diversion, I thought. As we kick-started the bikes, Cal stood outside his doorway with a Lucky Strike in his hand. "Careful," was all he said.

Only a short time later did it occur to me that if even one of these desperadoes had half a brain, they would see right through the ruse and immediately head south on Highway 77— really the only way to the border town of Brownsville. Too late for second-guessing now. We were headed for Mexico!

A short while later on the outskirts of Corpus Christi, we came to a gas station with a homemade sign: "Last Gas for Sixty Miles." We were lucky. It was closing up for the night, and we caught it just in time. After this, the highway was a desolate stretch through the King Ranch, the largest ranch in the country, we were told, and we wouldn't make it without a fill-up. We had no choice but to fill the tanks, all the while looking nervously behind us, hoping the delay wouldn't prove costly.

Back on the highway, it was pedal to the metal. I remember thinking that if a group of cars came up behind us, we could always leave the highway and try to lose them in the mesquite, sincerely hoping it wouldn't be necessary. The good news was that the highway was absolutely deserted, allowing us to hurtle along at a steady seventy, a speed we normally only did in short bursts. The bad news? The highway was absolutely deserted. So empty and desolate that if our pursuers—and we were still convinced they were on our tail—caught up with us, we would be at their mercy (to which, so far, they seemed to show no inclination). We were not ensconced in the secure exoskeleton of an automobile. We were on motorcycles, exposed to the open air. And wasn't that supposed to mean freedom? In our present situation, it imparted a heightened sense of vulnerability. We were feeling it. I recall looking up at the half moon and thinking that there would be some light if we had to shut off our headlights and execute an off-road disappearing act.

Our headlights would occasionally pick up deer browsing alongside the highway. First, the greenish glow of their eyes, then their vague ungulate shapes. "Stay right where you are!" I implored them telepathically. They cooperated...for now, at least. Nevertheless, my mind darkened with nebulous premonitions of violent impact if one of them should unexpectedly dart across the highway just as I was in the right place at the wrong time. That thought was saddled with the irony of my possible demise at the hands of a 150-pound trophy buck while fleeing the bats and chains of a band of Texas miscreants. And how could you blame the deer? He and his ilk had been playing in the mesquite long before humans redefined the landscape with ribbons of asphalt and speeding internal-combustion conveyances.

And not just deer. Once or twice, an armadillo scampered across the pavement. What if your bike struck one of those armored quadrupeds? These primitive creatures remain basically unchanged from ancestors that roamed earth eighty million years ago, but they too have had to contend with fast-moving motor vehicles only in the past half century. And, judging by their many carcasses littering southern highways, they hadn't yet adapted. What did Darwin's *On the Origin of Species* have to say about that? Talk about irony. (And while you're at it, have fun with my epitaph. What possibly rhymes with *armadillo?*) And was it even a premonition I was experiencing? Can it be said that you have a premonition if the event never actually takes place? So maybe it was just a thought, albeit an ominous one. And why was I racing through the King Ranch in the middle of the night, playing pseudosemantic, philosophical mind games? The things that went on in my head!

I told Paddy to keep an eye out behind us. The damp, coastal air was tightening my traumatized neck muscles, and a stabbing pain had set in, making it impossible to turn my head even a few degrees. That part of me had always been problematic, but it had never felt worse than on this night.

When we arrived at the border checkpoint, it was midnight and there was no delay. Crossing into Mexico back then was fairly easy. I think we showed the guards our driver's licenses (hoping no suspicions would be aroused by a Vermonter and an Arkansas resident with New York accents) and answered a few questions about the purpose of our visit. "*Tourismos*," I said (leaving out the part about needing to put an international boundary between us and three carloads of reprobates), and how long we planned to stay. We were waved on.

We had escaped to Mexico!

Wired as we were, we decided not to stay near the border town of Matamoros but to continue on farther—the farther the better!

As we rode off into the warm Mexican night, my thoughts turned to the fräuleins. We had said our auf Wiedersehens just a few hours back. This translates to "until we meet again," and we did indeed have a date for the next evening, did we not? A date that was not going to happen. They would be standing there, looking for us, and we would be over a hundred miles away with no means of communicating. What would they be thinking? How long would they stand and wait before giving up, saying that those nice *Amerikan* boys were maybe not so nice after all, standing them up like that! How I wished they could know what had really happened. "Don't be angry." I silently implored. "And don't be worried either. I think it's going to be OK. We're in Mexico now!"

Escape from the USA

I've told this story many times over the years, and invariably I'm asked why we didn't just call the police. It's a fair-enough question. The answer, in all its simplicity and complexity:

we were two young guys
from the North
traveling through the South
on Harley-Davidsons
in 1966!

The Kindness of Strangers

Each journey has a halfway point. The event can be determined chronologically or by distance traveled. It can be prescheduled so you're aware of it when it takes place, or sometimes it just happens, which is more or less the way it was when we got to Monterrey, Mexico. We had entered that industrial city very late at night after a day of sightseeing in the surrounding countryside, amazed to find the place bustling with people, even small children, at an hour when even most of Manhattan would be relatively empty. The tagline "the city that never sleeps" really belongs to this capital of the state of Nuevo Leon—or so it seemed to us.

Here, we agreed that maybe the time had come to start heading home. Maybe not directly. We'd meander every now and then and still get to experience the joy of getting slightly lost, but we had already been on the road for almost a month, had seen a lot, and certainly had been very lucky when you consider the close call in Corpus Christi. The motorcycles were also a concern. Not that they were giving us problems; they were actually running quite well. But they weren't exactly new, after all, and you never know. The deeper we traveled into Mexico, the longer the return trip back to New York. And money was another issue: frugal as we were, we had each already wired home for some additional funds through Western Union. My savings account was dwindling, and I would need something when I started graduate school in late August.

Escape from the USA

Certainly it wasn't the people of Mexico who influenced our decision to head back. They were, without exception, friendly and welcoming. The young men especially seemed always eager to know more about our travel adventures, and the motorcycles would attract curious crowds wherever we stopped.

This was also our first experience with Mexican food, which at this time was relatively unknown in New York—or, for that matter, in the entire East. Taco Bell came on the scene much later. And "Tex-Mex"? That term didn't even exist in the '60s. In keeping with my "when in Rome" philosophy, I was eager to sample Mexican food (not that there was any choice, since that was all there was), and in the trial-and-error process learned soon enough to request "*no picante*" when ordering! Of course, it wasn't called "Mexican" food down here anymore than people in China would say, "Let's have Chinese tonight," or a Frenchman would order "French fries." And in Hamburg, Germany, there's no such thing as a hamburger—you get the point. Nor could it be said that we threw caution to the winds when partaking of tacos and burritos, since I don't recall knowing anything about "Montezuma's revenge." We just ordered what looked good. And, for the most part, it was—and cheap!

The Mexican night was dark. Rural areas had minimal electric service, and several times, we pulled over and shut off the headlights just to look up and actually be able to see the Milky Way galaxy that we are part of but could never see back in New York. On one occasion, we were seeking a place to put our sleeping bags down after a day of sightseeing. Field of vision was limited to the range of our headlights and a couple of puny flashlights we had packed for the trip—none of which provided great illumination—but we did manage to find a nice, flat area surrounded by trees and far enough from the road so we wouldn't be disturbed by traffic. Our gas tanks were running low, and we didn't want to take any chances. Air mattresses inflated, we crawled into sleeping bags and waited for Morpheus to pay his nightly visit, which didn't take long.

It wasn't daylight or the warmth of the rising sun that awakened us but the clucking of a flock of hens, some of which were bold enough to perch on top of us. Rubbing our eyes, we immediately became aware of two boys, brothers apparently, maybe ten or twelve years old (it was hard for us to tell) standing quietly not ten feet away. And for how long they had been standing there politely, waiting for us to open our eyes, I have no idea. Seeing that we were awake, the older boy asked in more than passable English, "Do you want to see a beautiful waterfall? My brother and I will take you there."

A bit startled at first—after all, we thought we had found a secluded spot to lay our heads last night—we got over our surprise and then stood up and looked around. Not fifty yards away was a modest adobe farmhouse. It was then that we realized we had slept in their barnyard! Standing beside the dwelling was an adult couple, the parents no doubt, who waved when we looked their way and in doing so immediately allayed any concerns we had about trespassing. We discreetly put on dungarees and T-shirts. As the parents approached us, we were hoping they understood some English and immediately apologized for staying overnight on their property. "It had been very late and dark, and we didn't realize…"

"No problem," they assured us.

"That waterfall the boys want to show us…can we wash up there?" I took a cake of soap from my ditty bag and pantomimed the act of bathing.

"No problem."

"Give us a minute to get our clothes together, and a towel. And, by the way, how far away is the waterfall?"

Escape from the USA

"No problem," they assured us.

In a hushed voice, I said to Paddy, "I think maybe they understand English better than they speak."

"No problem," he hushed back.

So, off we went with the boys in the lead, who actually could speak English fairly well, down a well-used footpath, and after about a mile, we heard the sound of the waterfall before we actually saw it. The setting was truly idyllic. But there was evidence that it was frequented by the locals. We gave the boys a few pesos and told them we would see them back at the house.

The creek that fed the waterfall was actually not that big. Narrow enough, in fact, that with a running start, you could probably jump across. But the vertical drop was at least thirty feet before hitting a shallow splash pool, and with the pressure of that water cascading down on our backs came a pleasurable tingling sensation. Overused as the word is, "sensational" best describes it. *Sensational*, as in engaging of the senses: heightening them in such a way that all is blocked out except for the moment at hand. Surely such an experience could not be sustained for long, but at that moment, it was transcendent, even sublime. The temperature was perfect and actually therapeutic for my cervical vertebrae, which, by the way, were feeling much better, although not 100 percent (thanks for asking). What a way to start the day!

Upon our return to the motorcycles a half hour later, the boys, with big smiles, asked, "Did you like?"

The look on our refreshed faces should have been a dead giveaway, but we let them know how beautiful it had been and thanked them very much with an additional couple of pesos. The parents had now joined the boys to see us off, and we thanked them as well. "*Gracias, gracias!* Thank you for everything. Thank you for letting us sleep in your yard...thank you for the waterfall...and thank you for *hospitalidad!*"

"No problem," they replied in unison.

We secured our luggage and started the bikes, which put more big smiles on the boys' faces. The older one asked, "You can go very, very fast?"

"Adios!" I managed one final word of Spanish as we slowly rode off before getting onto the pavement and speeding away, much to the delight of the brothers. Looking back over our shoulders, we could see them still waving as we headed down the highway.

The next day, a troubling noise started coming from Paddy's bike. I didn't hear it at first, but every Harley rider is in touch with his machine, and he knew something was amiss. Fortunately, having experienced the problem once before, he diagnosed it immediately: the primary chain sprocket was loosening. Unfortunately, the fix required a fairly large wrench of a size that we didn't carry with us. We had tools, but only enough for minor repairs and maintenance.

We were stopped at a small roadside rest area, pondering our dilemma, when a fellow in a green 1957 Chevy pickup pulled up. Out of his window, he asked—in English—did we need any help? We explained the problem to him.

"I've got a tool chest at home. Stay here, and I'll go and get it…I'm only about ten miles away. I'm Raul, by the way."

A half hour later, Raul returned with a bright-red tool chest in the back of his truck. Paddy removed the primary chain cover, and within ten minutes, the problem was fixed. We offered to pay Raul for his services, but he wouldn't take a dime.

"Can we at least pay you for the gas you used going back and forth to your house?"

"No way," he insisted. "Just remember there are good people here in Mexico!"

THE EPISODE

It was late afternoon on the first day of August. It seemed like we had just become acclimated to the hot Mexican sun, and now we would be returning to more hot sun: the Texas variety. The abiding theme of our thoughts now had to do with the trip home. The reentry point would be Laredo, considerably farther up the Rio Grande from where we had left Texas some five days ago, and we were hoping the locals in this part of the state were capable of more hospitality than their brothers in Corpus Christi. About fifty miles south of Laredo on Route 85, we came to an active little truck stop and decided to refuel and grab a bite to eat.

The cantina was bustling with men. As usual, the motorcycles attracted attention, and as soon as the kickstands were down, a few fellows approached us. Small talk ensued of the where you from, where you been, and where you goin' variety, but this didn't last long. One of the guys, a Mexican truck driver in his early thirties, asked us in excellent English, "You heard what happened up in Austin today, didn't you?

Something about the gravitas of his question concerned me. "No...what happened?"

"You didn't hear?" he asked, this time with incredulity in his voice.

"We haven't heard any news for a while. You know how it is riding..."

"It's all over the news. A fellow in Austin with a sniper rifle climbed up into a tower and shot a bunch of people for no reason. I think there's over fifteen dead."

My immediate reaction was angry disbelief and probably some physical reaction, judging from the way the fellow looked at me. I looked back at him, especially at his eyes. They always say you can tell by the eyes. I was searching for some indication that he was just a disturbed person messing with our heads and this was his idea of a sick joke. I made the quick assessment that this guy was not crazy. "What are you saying?" I demanded.

"I'm telling you what's all over the news. A guy with a gun went up to the top of a tower in Austin and started shooting people…for no reason. Go inside." He pointed in the direction of the cantina, "Everyone's watching it on TV!"

In the next few weeks, the world would learn more about the "guy with a gun." By all outward appearances, Charles Whitman had been the All-American boy. Good-looking, with a blond crew cut, he had a reputation for being a "well-mannered child who never lost his temper" and apparently still didn't while methodically gunning down over fifty people, sixteen of whom subsequently died of their wounds. As a youngster, Whitman had been an altar boy (I don't mean this metaphorically) in his local Catholic church and later, at the age of thirteen, earned the rank of Eagle Scout in his hometown of Lake Worth, Florida. After graduating from high school with honors, he had joined the Marine Corps and earned a Sharpshooter's Badge and Good Conduct Medal. But then things started going downhill in his military career and personal life, and it all came to a culmination in that tower at the University of Texas.

I remind you that this was 1966. I was twenty-two years old, and nothing like this had ever happened in my lifetime.

Escape from the USA

Paddy and I entered the cantina, and sure enough, a bunch of guys were sitting around drinking beer, their eyes fixed on a small black-and-white TV behind the counter. The announcer was speaking in loud, rapid-fire Spanish that made no sense to us, but the imagery on the grainy screen made it clear that something chaotic was still unfolding. Men at the bar were speaking in hushed tones, shaking their heads. My disbelief was slowly replaced by a strange feeling of despair tinged with anger. But mostly despair.

All sorts of crazy thoughts and emotions were scrambling through my head. What kind of country were we returning to? Kennedy's assassination had taken place less than three years earlier, and that event had shocked the world. I remember thinking at the time that that sort of thing didn't happen in the United States—maybe in banana republics. We Americans were civilized. But even that was different, being an assassination with political motives. This Austin thing was a random, senseless thing involving innocent victims! What kind of crazy, fucked-up country was the United States? Or was it just Texas? Less than a week earlier, after all, we had fled the state, fearing for our own lives.

Of course, in the rest of the twentieth century and into the twenty-first, mass shootings would take place, and with increasing frequency.

And there is another form of tragedy that has affected all Americans, especially those too young to have any memory of the pre-1966 years. I'm referring to a sense of resignation and even acceptance regarding these crazy outbursts of random violence against innocents that have become almost routine in the course of our lives.

I remember the pall that descended over us in the aftermath of the University of Texas episode and verbalizing something about the terrible coincidence that we were to travel through Austin tomorrow, the site of the killings. What had brought us to this crazy, fucked-up state of Texas at this point in history? I also recall thinking that as weird and crazy it was to actually be so close to the location of the massacre at this time, one thing was certain: that precisely because it was so horrific—surreal, actually—we would *never again* see the likes of such an episode in our lives! Little did we realize then that such events would happen again and again and again and again and again and again and again and again and again and again and again and again and again and again and Again and Again and Again and Again...

Herein lies the real tragedy. In the aftermath of each subsequent American mass shooting, no one ever asks *if* another mass killing will ever take place. Americans know by now that it's simply a matter of *where* and *when*.

Postscript

Grant Duwe, a criminologist with the Minnesota Department of Corrections, has studied more than thirteen hundred mass murders that took place from 1900 to 2013. There were few before the 1960s.

In the *New York Times* on October 4, 2015, Dr. Duwe was quoted as saying that "the *episode* that some academics view as having introduced the nation to the idea of mass murder in a public space happened in 1966, when Charles Whitman climbed a tower at the University of Texas in Austin and killed 16 people."

The Way Home

We had already agreed that the time had come to head home. We had been on the road for over a month, and quite frankly, the adventure had run its course. Or ride home included less meandering and more straight runs on portions of newly completed interstate highways. In 1966, The Dwight D. Eisenhower Interstate Highway System had about half the mileage that it does today (2017). It was during the '60s that most these miles of pristine new pavement were added for our riding pleasure.

But since we were already in the central portion of the country, the return route was to include a visit to Champaign-Urbana, Illinois, where in less than a month I would be attending graduate school for my master's degree. I thought it would be a good idea to spend a little time there by way of a self-guided orientation. Paddy and I also thought about driving a little farther north to Milwaukee to visit the Harley-Davidson factory. Maybe they'd like to hear about our exploits. But perhaps we were getting a little ahead of ourselves, for at the moment, we were entering the border town of Laredo, Texas, and a pall still hung above us as a result of the Austin shooting.

Whenever we stopped for gas or food or for any reason, conversations around us had to do with the incident at the University of Texas. Our route north would take us through the city of Austin, but we had already decided against spending time there, not wishing to appear like morbid gawkers at a disaster scene.

But San Antonio was also on the route home, and it would be unthinkable to not pay a visit to the Alamo! Back in 1954, at the age of eleven, I had been captivated, along with much of America's youth, by the *Davy Crockett* Disney miniseries on TV (to this day, I can still sing "The Ballad of Davy Crockett"—the first verse, anyway—from memory). We all know the story: Crockett, Jim Bowie, and Colonel William Travis, along with 180 valiant defenders, had died fighting to the last man against the perfidious Mexicans.

I remember thinking how small the mission appeared in contrast to my own Disneyfied expectations, and the brief tour did nothing to alter my understanding of the event as recalled from sitting in front of the TV as a youngster. Later in life, I learned a more nuanced view of the Texans' rebellion against Mexico and that the events of the Alamo might not have gone down quite like the Disney version: that maybe maintaining the right of Texans to hold slaves had been part of the motivation for its secession from Mexico. OK, maybe the Disney version was more "appropriate" for eleven-year-olds in the 1950s, and it simply became one more on a growing list of American-history facts that I had to relearn later. When Sam Cooke sang, "Don't know much about history," he was actually in good company with American schoolchildren of the 1950s.

After the Alamo, I was tooling down a country highway, mesmerized by the pastoral landscape in the vicinity of Waco, when I became aware that my companion was no longer following me—and, in fact, was nowhere to be seen or heard. How long I had been riding solo, I had no idea. Seeing a small roadside rest with picnic benches, I decided to pull over and wait, and if necessary, turn back to see if he had run into any problems.

Escape from the USA

It turned out I didn't have to do this, which was a good thing, seeing as how we had recently become interested in actually making progress in our travels. I heard his bike well before he pulled into the rest area wearing a big shit-eating grin, proudly pointing at the enormous watermelon bungeed on top of his luggage.

After getting laughter under control, we proceeded to attack the purloined prize, its illicit provenance clearly adding to our dining pleasure—if it could actually be called that, since we had decided that this beauty deserved to be consumed "barbarian-style" in our own parody of the "sword and sandal" movie genre popular during the '50s and '60s. Despite further spasms of raucous laughter, the giant melon was devoured in an astonishingly short time.

As we relaxed at the bench, surveying the remains of the watermelon carcass scattered about us, I remarked in perfect Barbarianese, "It is good to pillage; is it not, my friend?"

Paddy picked up on the cue. "But, alas, would there not be greater satisfaction with grog to quench our thirst and wenches at our side?"

"Then, by the Hammer of Thor, we shall descend upon the next village, and we shall have our wenches!"

More laughter.

Feigning gravitas, I then brought attention to the possibility of being apprehended by the police. "You realize, I hope, that we're still in the state of Texas…"

"So what if we're in Texas?"

"For one thing, they still shoot cattle rustlers and horse thieves. You watch Westerns…"

"It's a friggin' *watermelon*!"

"Yeah, so that makes us watermelon rustlers!"

"I didn't lasso it. I picked it out of a watermelon patch!"

"It's still against the law…criminal mischief, petty larceny, maybe even pillaging and plundering. The evidence is all over our faces. What if we get busted?"

"What do you mean, 'we,' *kemosabe*?" (This was the punch line of a popular joke of the time.)

More laughter.

"This is true. It will be only you that gets busted. I'm only eating the watermelon. You actually pillaged it!"

"It's called aiding and abetting—also illegal!"

"Aiding and abetting a watermelon rustler?"

"Totally illegal in the state of Texas!"

More laughter.

And there were more inane watermelon scenarios, even some risqué ones I'll leave to the imagination of the reader. After the Austin shooting, we sorely needed levity. You can't be riding a motorcycle bummed out like that. It's just wrong.

The Hill Climb

We finally arrived in Champaign-Urbana, and I got the opportunity to spend a few hours orienting myself to the campus. Paddy stayed with me for a while before we agreed to split up and meet a few hours later so I could continue exploring the neighborhood on my own. I found the house owned by Miss van Deventer, who would have a room waiting for me upon my return in less than a month. It was an old colonial-style home on a street lined with American elms and well within walking distance of the campus. I had already sent her a deposit sight unseen, so I was relieved to see that it appeared to meet my needs. I saw no need to introduce myself to my landlady. She might have had difficulty with the idea of renting her room to a Harley-riding graduate student. When I showed up a month later, it would be on four wheels: those of the 1961 Volkswagen Beetle that Dad had given me for graduation.

I would discover that Miss van Deventer was actually eighty-two years old and insisted upon her "Miss." I would also learn that she was a card-carrying DAR member, had a fascination with flags of all nations past and present, and was still in the habit of referring to black people as "darkies" ("Back when I was a teacher in Springfield, I was really strict with those little 'darkies!'"). But back to the story.

Our original plan had been to continue north and visit the Harley factory in Milwaukee. That plan was shelved. We were both antsy about getting home. Paddy did, however, fly out to Illinois during my winter break a few months later, so we did eventually make that trip to Milwaukee for the factory tour as part of his birthday celebration.

If you lived in the eastern part of the States in the mid-1960s, you might remember that much of the region was locked into a serious drought: bad for farmers and lawn-obsessed suburbanites but great for motorcyclists. On our road trip, for example, we had been away for over a month and encountered only one brief rainstorm (the one when we ducked into the hayloft for refuge). That was it as far as rain. So dry was it that we didn't even bother to pack rain gear.

So when the skies started to darken and the first few drops began to fall in the late afternoon while we cruised on I-70 somewhere in eastern Ohio, we were not only surprised and unprepared but a bit peeved as well, because we had planned to spend that evening somewhere in western Pennsylvania, and that now seemed unlikely. Soon the raindrops got heavier, forcing us to take shelter under a brand-new overpass. Not a moment too soon, as it turned out. As we ducked under, rain started coming down in buckets. The little red Emerson transistor radio came out from the luggage, and we finally got a weather report, which confirmed what we had already kind of figured out for ourselves: rain all night, ending sometime in the morning, heavy at times. Neither of us had the stomach for riding in a drenching rain to the nearest motel, wherever that was, and quite honestly, money was starting to be a factor (meaning the lack thereof). We scoped out the confines of the underpass and discovered a cozy little niche at the top of the embankment, to this day a popular place of refuge for thousands of America's homeless. Any port in a storm, goes the saying.

Escape from the USA

The deluge made the decision easy: we were in for the night. With our bikes parked far as possible off the road, we scampered up the incline with air mattresses and sleeping bags and found a nice firm, flat surface—good for your back, ask any orthopedist. Fortunately, we had some snacks packed away for just such an emergency, so we didn't go to bed hungry. You would think that the roar of diesels motoring east on I-70 and the intermittent thumping of vehicles overhead would have made for a less-than-ideal night's slumber, and you would be 100 percent correct. But it's surprising how adaptable the human species can be. Before falling asleep, an earworm entered: the harmonies of Brian Wilson and the Beach Boys singing "In My Room." A serenade, or some sort of sick joke?

Next morning, we awakened well before sunrise and half walked, half slid our way down the embankment for a weather check. Good news! Precipitation had ceased as promised, and already a soft glow lit up the eastern sky. In a now-familiar ritual, we deflated the air mattresses, rolled up the sleeping bags, and were soon on the road again. If the weather cooperated, we would be back in Queens sometime the following day.

Later that day, riding through on a country road in western Pennsylvania, we came upon a dirt track going straight up the side of a hill in pastureland...a motorcycle hill climb! We stopped and took a nice, long look. Then we looked at each other and smiled. Why not?

Why not? Actually, for a lot of reasons. For one thing, we were on Harley hogs, not hill-climbing machines. Hill races consist of steep inclines on dirt tracks marked off in segments. The rider covering the greatest distance before the inevitable flip is the winner. Not surprisingly, specialized machines are used for this. Because of the steep angles, the bike frames are extended in the rear so that the rider (and therefore the center of gravity) is moved forward to postpone the inevitable back flip—it's just plain physics. The other difference is the tethered kill switch needed to "kill" the engine when rider and bike part company. These machines also have knobby tires for traction (our tires were barely street legal!), and competitors are suited up in protective garments and helmets. The fact that Paddy and I had decided to go for it despite the lack of every one of these prerequisites was not exactly a testament to our intelligence.

We unbungeed our gear, and to make it interesting, threw in a little wager: whoever went farthest would get treated to dinner by the loser. I won the coin toss and decided to go second, so Paddy mounted his bike and positioned himself about thirty feet from the base of the hill. Looking up with a big ol' smile, he launched himself off the starting line. Once he hit the incline, the first twenty to thirty yards were fairly manageable. Then the ascent suddenly got steeper, and his struggles began. He made an effort to shift his center of gravity forward by standing on the footpads and leaning over the handlebars, and for a few seconds, the strategy worked. The big bike still managed to make forward progress, haltingly, with the poorly treaded rear wheel struggling to maintain grip, spinning more than biting and spewing dirt and stones against which I had to shield my face even though he was far beyond where I stood.

Then came the inevitable. The front wheel slowly left the ground, leaving Paddy suspended in the air for a few moments, looking down over his left shoulder. The hill climb was about over for him. But how would it end? With him bailing out and throwing himself away from the machine, maybe hitting the ground with a bone-sparing roll? Or with him tightening his grip and going down with the bike? In the next second, the bike went completely vertical, and the front fork jerked to the right, wrenching the handlebars from his hands. Although he had somehow managed to land on his feet, he hit the ground ludicrously running downhill, which we all know can't last long. So, out of self-preservation, he went into a barrel roll, finally winding up about thirty feet downhill from the bike, which rested supine, front wheel furiously spinning. Jumping to his feet, he half ran, half crawled back to his bike and, after a mighty struggle, managed to set it back up on its wheels.

Still fighting to hold his machine upright, he looked down at me and yelled, "Your turn!"

Not thinking his challenge warranted a direct reply, I yelled back, "Let's just mark the spot where you stopped going forward...I can already taste that steak!"—by way of a reminder that a wager had been made here.

"You can always back down and just buy me dinner anyhow, you cheap kraut!" (I took no offense. Political correctness hadn't yet been invented.)

"Sounds like you're worried..."

"Worried that I'll have to take you to the hospital!"

As foolish as this caper was starting to look, there was no way I was going to back down, especially after Paddy had already made an attempt. But I was a bit worried. My friend was a good rider, after all, and that hill had made him look silly. Wouldn't I be the bigger fool if I didn't learn anything from his experience? I could still back down. But I still was at a point in life where I had a lot to prove—and how I hated the sound of that expression: *back down!*

And there were other considerations. How about the obvious: we had already traveled about 4,000 miles, and this little hill was a mere 150 miles from home. What if I did get hurt and wound up in the emergency room? And what about the bikes themselves? They had held up remarkably well mechanically. What if the hill climb resulted in damage? More than inconvenient, it would be plain stupid. I recalled a decision-making strategy I had learned where you write out a statement and then list the pros and cons with a numerical value attached to each; you then simply add up the total value of each column. Had I used the strategy, even with a so-called fudge factor thrown in, the results would have been overwhelmingly in favor of ditching the idea. If only all life choices could be reduced to such simple paradigms! "Let's go mark the spot where you went down."

At our feet was a wooden stake that we picked up, and together we trudged up the hill. The spot where Paddy had stopped making forward progress was easy to locate because of the fresh rut dug out by the rear wheel when his bike went perpendicular.

Paddy negotiated. "Give me two more feet, this being the rear-wheel mark."

"Fair enough," I consented, confident that I could still best him, and stuck the stake into the ground a few feet off to the side of the track.

"Tell you what. I'll stay up here, off to the side," he suggested.

I thought about it for a few seconds. Why would he do that? Maybe to make faces and distract me or make me lose my concentration…I wouldn't put it past him. But, wanting to be a good sport, I said, "OK, no problemo. Just not too near the track…and don't be sticking anything into my spokes!" I pointed to a flat boulder about ten feet off to the right. "There's a good spot."

Paddy perched himself on the rock with his customary ear-to-ear smile while I half walked, half slid my way back down to my bike.

At the base of the hill, I shielded my eyes against the sun and looked up. My friend was now lying back, legs crossed with hands locked behind his head, making an effort to exude extreme confidence. Even at a distance, I could see the smug smile was still there.

I decided to attack the hill somewhat differently than Paddy. He had stayed in first gear throughout, and it seemed to me that the high engine revs had caused unnecessary wheel spinning, especially right at the start of the actual ascent. My plan was to get off to a smooth start and then hit second gear right before the actual incline, hoping this would give me greater initial speed, result in less wheel spin, and provide me with the momentum necessary to beat my friend.

I put the front tire on the starting line we had made, shifted into first, and revved the engine. Paddy had now gotten back up to a seated position, elbows on knees, his still-smiling face now resting in the palms of his hands. I released the clutch smoothly, not wanting to spin out at the start. The bike lurched forward, and in seconds, the revs were just right for the shift into second right before the start of the incline—perfect! When the incline began, revs were right where I wanted them and climbing. Sliding forward on the saddle, I leaned over the handlebars, and although the rear wheel was starting to spin and churn, I still had traction enough to maintain forward movement. That stake we had planted in the ground was still in front of me but coming up rapidly.

Out the corner of my eye, I could see Paddy up on his feet, arms raised. Was he signaling victory for himself? Just then, I felt a jolt in the front forks, and the wheel began lifting off the ground. I maintained a grip as long as I could as the front end lifted. But for me, the climb was over, and in a final effort, I swung the handlebars to the left, let go, and threw myself off the bike. I recall being airborne for a period that probably felt longer than it really was and then hitting the ground running before going into a series of head-over-heel rolls that ended in an abrupt halt just before a large boulder. I immediately looked up in the direction of my bike, which had come to a rest about twenty feet uphill.

An adrenaline-fueled sprint got me back to the bike at about the same time as Paddy, and together we wrestled it upright. But we almost immediately became aware of an acrid smell and saw smoke billowing out from under the gas tanks. The electrical system had shorted out because of the severe jostling of the spill. Remember, there were no fuse boxes on those machines! I had to disconnect the battery quickly, or else the whole wire harness would melt down. I jumped on the bike and got back down the hill as fast as I could. Frantically searching, I finally came up with a stout screwdriver that I used to pry the positive cable off its terminal, but I knew some serious cooking had taken place.

Using the tools at our disposal, we removed the gas tanks to inspect the damage. Yes, a number of wires had their insulation burnt off, but it could have been worse. The good news was that I had taken along a spool of electrical wire for the trip, along with a box of terminals. After patting myself on the back for my prescience, I proceeded to replace badly exposed wires and patch up the not-so-badly compromised wires with electrician's tape. A couple hours later, the job was done.

No question about it, we were guilty of motorcycle abuse. Before departing, we went over the machines carefully, making minor adjustments and repairs along the way. Then we started the engines and checked the oil and gas lines for problems, but to our relief, we found none. The bikes got a clean bill of health. With saddlebags fastened back in place and luggage rebungeed, we rode off into William Penn's sylvan landscape.

An hour or so down the road at about 6:00 p.m., we passed a road sign announcing our entrance into Allentown. Paddy pulled up alongside of me at a traffic light with a big shit-eating grin. Something was up. "Looks like a good place to stop for that steak you owe me." He was trying to act serious through that big-ass smile.

"Funny you should mention it, because I was just about to say, 'Don't worry, Paddy; I won't order the most expensive steak on the menu, seeing as how you're running low on cash and all and me being your best friend.'"

It suddenly occurred to us that in the dust and dirt of the hill climb and the calamity of the short circuit, neither of us had taken time to actually figure out who had "won." But we did stop in at a diner, where my friend ordered himself a hamburger. There was scrapple on the menu, and this being Pennsylvania, I ordered it.

"Scrapple…for dinner?" the waitress responded quizzically, making me think that maybe I had committed some sort of cultural faux pas.

"Yeah, scrapple…is that possible?"

"Anything's possible, hon!"

I paused for a moment to consider the possibility of some hidden message in her reply…and whether we actually had any money left to pay for the meal. "You won't get that in New York," I said to Paddy by way of justifying my menu selection.

"I know," he said. "When in Rome…"

At this point, our waitress was starting to look a bit perplexed, having probably taken thousands of scrapple orders without the fanfare she was now witnessing. I decided to cut to the chase. "Then I'll definitely order the scrapple!"

Our waitress cheerfully wrote down our order, turned around, and headed to the kitchen.

"Speaking of New York, what do you say we just keep riding and make it back home tonight?"

"It'll be late when we get in—so late that it'll probably be early. But that could be a good thing. No traffic."

"It's a plan!"

After we ate and finished our coffees, I broke a dollar bill and took the change to the phone booth outside the diner. Immediately after I dialed home, the operator requested "forty-five cents for the next three minutes, please."

"Hello?" It was my mom.

"Mom, it's Erv...how are you?"

"Ervie"—that's what she always called me—"where are you? Is everything all right?"

"Everything is great, Mom. We're in Pennsylvania. I should be home tonight, but it'll be late, maybe one o'clock in the morning. So leave the door unlocked. I don't have a key."

"One o'clock? Are you sure you're not too tired to drive?"

"Wide awake, Mom. Don't worry."

"You've been gone a long time. Dad is asleep on the couch. I don't want to wake him up."

"Don't worry, I'll be home soon, and then we can catch up."

Just then, the operator butted in. "Twenty-five cents for the next three minutes, please."

I thought I had a few seconds before getting cut off. "Mom... got to get going. What's for dinner tomorrow night? How about *Rouladen*?"

My request kind of hung in the air when I lost the connection. I hoped Mom had heard me. I had been thinking about a good home-cooked meal for a couple of days already, and now I could taste it.

Epilogue

No man ever steps in the same river twice, for it is not the same river and he is not the same man.

—Heraclitus

The odyssey had come to an end. And although we hadn't been expecting a ticker-tape parade, reentry into College Point was less than triumphant. You might even say it was anticlimactic. Well, it *was* after midnight. My family was there to greet us warmly, as was Paddy's, and the next day, friends dropped by to hear all about the trip. Of course, Mom had prepared the requested dinner for the following evening. Food is love, after all.

Let's not forget that motorcycles are dangerous, and the fact that we had returned intact was not completely taken for granted. But the full extent of our adventure would only unfold piece by piece in the months and years to follow—not just our story together, but also my own story and Paddy's own as well, since each person's experience is his and his alone.

We had accomplished something quite remarkable, covering five thousand miles on old rebuilt Harleys, and had passed a test of sorts, a test of physical and mental endurance. Let's face it, many people wouldn't think of driving that distance even in a comfortable air-conditioned automobile. Though motorcycles have become more popular than ever these days, I often see guys trailering bikes to wherever it is they're going. So much for the adage, "It's not the destination, it's the *journey*!" Perhaps its point was always more metaphysical anyway.

And what about the motorcycle? Frankly, I had already been thinking about selling my beloved 1949 Harley-Davidson for some time. I certainly couldn't take it out to Illinois with me, and I did need the money. But a small part of me kept saying, "Keep the bike." There would have been no problem storing it in my parents' two-car garage. There was plenty of room. Then, maybe when I returned to New York (I had assumed I would eventually return and find a job there), I could go riding on days off and weekends. But somewhere on the return trip from Mexico, I had made up my mind.

I had completely built that old Harley with the help of friends and had gotten several good years out of it and at least one memorable adventure. And even though I saw other riders go down with catastrophic injuries, I had escaped relatively unscathed myself. Some might even say I had "cheated death." Maybe it was time to quit while still ahead.

Furthermore, I reasoned, the bike was still running flawlessly, so maybe the timing was right. I decided to test the waters and put word out that the bike was on the market for $800 (about double what I had paid for it three years prior). Considering the time and money I had put into rebuilding it, I would be pretty much breaking even.

Within a few days, a neighborhood guy known as Naples Joe (which is what he called himself, since he had, in fact, come to America from Italy as a boy) got in touch. "Erv, I hear you're selling your bike."

"I am, Joe. Hate to get rid of it, but I could use the money." (All true.) "I'm asking eight hundred."

Joe knew my bike and knew about all the restoration work I had done. We had even gone riding together a few times, Joe with his Sportster. "You selling the Sportster, Joe?"

"No, I'm keeping it, but my wife said, 'You want me to go riding with you? You get something more comfortable!'"

"You know the machine. I wouldn't sell it to you if I didn't think it was a good bike."

Joe also knew that Paddy and I had just returned from Mexico. "How did it run on your trip?"

"Perfect. No problems at all." I left out the part about the self-inflicted problems resulting from the hill climb. "In fact, take it out for the day. And if you still want it, it's yours."

He did, and later that night, he came back with a big smile and eight one-hundred-dollar bills. I signed the registration over to him on the spot. "I always like your bike, Erv!" he said in his faint Italian accent.

Postscript

The following summer, I came back from Illinois to stay with my parents in College Point for a while until I found an apartment. I had a teaching job lined up in Roslyn, Long Island, that would begin in September. In the meantime, I was working with our neighbor, Arturo, a plasterer by trade. He'd gotten me a job as a hod carrier at a new apartment building being constructed in the Sheepshead Bay section of Brooklyn. Arturo and I would drive early every morning on the Belt Parkway from Queens to Brooklyn to the jobsite where I performed the backbreaking work. But it paid well, so I was appreciative—and tired—at the end of each day.

And was that ever a hot summer! One afternoon on the way home, traffic on the Belt came to an abrupt standstill, and everyone got out of their cars and just sat on the grass alongside the parkway for several hours. What had happened was that a drawbridge had opened to allow a boat through, and expansion resulting from the ninety-eight-degree temperature then prevented the bridge from closing. Vessels from the fire department's marine division were summoned to hose down the bridge until it contracted enough to reclose—an adventure we could have done without!

So, exhausted from that long, hot day, I was sitting out on the front porch of my parents' house with my dad, enjoying postprandial conversation, when I heard a motorcycle approaching from the direction of Fourteenth Avenue. The sound was a familiar one. Looking up, I saw Joe rounding the corner on my old bike. He had always been a sharp dresser with sartorial tastes bordering on the flamboyant, maybe even the eccentric, like often wearing a Gary Cooper style cravat around his neck. So his shiny, new black-leather motorcycle jacket, one of those fancy ones with fringes, didn't really surprise me. But seeing the old bike again was like meeting up with an old friend I had lost touch with. The first thing you notice upon such encounter are the changes, and in this case, the bike looked and actually sounded better, I had to admit, than when he had driven off with it a year earlier. For one thing, Joe had obviously rechromed the parts that I had done on the cheap a few years back—what a difference that alone made!

"I love this bike!" were his first words before going on to describe the details of the work he had done over the past year, and I took the opportunity to compliment him. "Remember you always had that one valve that was a bit noisy?" he asked.

"Sure, I do...it drove me crazy. No matter how often I adjusted it, it still made that noise. So after a while, I just gave up and learned to live with it. How'd you fix it, Joe?"

"All I did was replace the one front pushrod with a slightly longer one...as easy as that."

"How come I never figured that out?" I wondered, confessing to the limits of my mechanical skills.

"I talk to the right people." He laughed. "You want to take it for a ride?"

"Thanks, but I don't think so."

We continued to shoot the breeze for a few minute before Joe finally rode off, with me staring at the departing exhaust pipes. I didn't know it then, but that would be the last time I ever saw the bike. Strangely, seeing my old bike again had aroused no desire on my part for another motorcycle. That chapter of my life was behind me, and I knew it.

My life has been rich with experiences and memories I will forever cherish. Some say that our lives are, in fact, the sum of these memories. If so, what do my memories of that singular, wonderful, and wacky journey during the summer of '66 say about me?

Every so often, I'm asked if I ever think about another motorcycle journey. Certainly the old wanderlust is still very much alive in me, and each year, some sort of travel is on my agenda. Even while writing this story, I'm in the midst of planning an adventure that just might turn out to be the best ever. But another motorcycle trip? The one we took back in the '60s is a tough act to follow. There's always a risk in trying to recapture such an experience, because what you're really trying to relive is a *feeling*, and I'm not willing to take that risk. It's best to heed the words of that wise, old Greek philosopher: "No man ever steps in the same river twice, for it not the same river and he is not the same man."

About the Author

Erwin "Erv" Krause is a semiretired educator. He is forever enthralled by travel and actively seeks all the world's unusual corners with his partner and travel companion, Lois. When not on his latest adventure, Krause splits his time between Long Island, New York, and Fort Myers, Florida.

For Your Listening Pleasure: Songs mentioned (or just alluded to) in Escape

1.	"Born to Be Wild"	Steppenwolf	1968
2.	"Urge for Going"	Joni Mitchell	1966
3.	"My Prayer"	The Platters	1956
4.	"Come Go with Me"	The Del Vikings	1957
5.	"Why Do Fools Fall in Love"	Frankie Lymon and the Teenagers	1956
6.	"School Days"	Chuck Berry	1957
7.	"Tutti Frutti"	Little Richard	1955
8.	"Love Me Tender"	Elvis Presley	1956
9.	"Sincerely"	The Moonglows	1955
10.	"Rockin' Robin"	Bobby Day	1957
11.	"Ballad of the Green Berets"	Sgt. Barry Sadler	1966
12.	"Eve of Destruction"	Barry McGuire	1965
13.	"1952 Vincent Black Lightning"	Richard Thompson	1991
14.	"Leader of the Pack"	The Shangri-Las	1964
15.	"Unknown Legend"	Neil Young	1992
16.	"The 59th Street Bridge Song" ("Feelin' Groovy")	Simon and Garfunkel	1966
17.	"Wild Thing"	The Wild Ones	1965
18.	"A Groovy Kind of Love"	The Mindbenders	1966
19.	"America"	Simon and Garfunkel	1968
20.	"The Times They Are A-Changin'"	Bob Dylan	1964
21.	"American Pie"	Don McLean	1971
22.	"Maybellene"	Chuck Berry	1955
23.	"Georgia on My Mind"	Ray Charles	1960
24.	"Midnight Train to Georgia"	Gladys Knight and the Pips	1973
25.	"Stand by Your Man"	Tammy Wynette	1968

26. "California Girls"	The Beach Boys	1965
27. "Rocky Raccoon"	The Beatles	1968
28. "Strange Fruit"	Billie Holiday	1939
29. "Crazy"	Patsy Cline	1962
30. "I Can't Stop Loving You"	Ray Charles	1962
31. "Mack the Knife"	Bobby Darin	1959
32. "The Ballad of Davy Crockett"	Fess Parker	1955
33. "(What a) Wonderful World"	Sam Cooke	1960
34. "In My Room"	The Beach Boys	1963

Made in the USA
Middletown, DE
26 February 2018